Meeting Regional STEMM Workforce Needs in the Wake of COVID-19

PROCEEDINGS OF A VIRTUAL WORKSHOP SERIES

Lida Beninson and Joe Alper, *Rapporteurs*

Board on Higher Education and Workforce

Policy and Global Affairs

The National Academies of
SCIENCES · ENGINEERING · MEDICINE

THE NATIONAL ACADEMIES PRESS
Washington, DC
www.nap.edu

THE NATIONAL ACADEMIES PRESS 500 Fifth Street, NW Washington, DC 20001

This activity was supported by contract No. 2005-1112031 between the Lumina Foundation and the National Academy of Sciences. Any opinions, findings, conclusions, or recommendations expressed in this publication do not necessarily reflect the views of any organization or agency that provided support for the project.

International Standard Book Number-13: 978-0-309-25628-5
International Standard Book Number-10: 0-309-25628-3
Digital Object Identifier: https://doi.org/10.17226/26049

Additional copies of this publication are available from the National Academies Press, 500 Fifth Street, NW, Keck 360, Washington, DC 20001; (800) 624-6242 or (202) 334-3313; http://www.nap.edu.

Printed in the United States of America

Suggested citation: National Academies of Sciences, Engineering, and Medicine. 2021. *Meeting Regional STEMM Workforce Needs in the Wake of COVID-19: Proceedings of a Virtual Workshop Series.* Washington, DC: The National Academies Press. https://doi.org/10.17226/26049.

The National Academies of

SCIENCES · ENGINEERING · MEDICINE

The **National Academy of Sciences** was established in 1863 by an Act of Congress, signed by President Lincoln, as a private, nongovernmental institution to advise the nation on issues related to science and technology. Members are elected by their peers for outstanding contributions to research. Dr. Marcia McNutt is president.

The **National Academy of Engineering** was established in 1964 under the charter of the National Academy of Sciences to bring the practices of engineering to advising the nation. Members are elected by their peers for extraordinary contributions to engineering. Dr. John L. Anderson is president.

The **National Academy of Medicine** (formerly the Institute of Medicine) was established in 1970 under the charter of the National Academy of Sciences to advise the nation on medical and health issues. Members are elected by their peers for distinguished contributions to medicine and health. Dr. Victor J. Dzau is president.

The three Academies work together as the **National Academies of Sciences, Engineering, and Medicine** to provide independent, objective analysis and advice to the nation and conduct other activities to solve complex problems and inform public policy decisions. The National Academies also encourage education and research, recognize outstanding contributions to knowledge, and increase public understanding in matters of science, engineering, and medicine.

Learn more about the National Academies of Sciences, Engineering, and Medicine at **www.nationalacademies.org.**

The National Academies of

SCIENCES · ENGINEERING · MEDICINE

PLANNING COMMITTEE FOR A VIRTUAL WORKSHOP ON ADVANCE CARE PLANNING: CHALLENGES AND OPPORTUNITIES[1]

JEFF BINGAMAN (*Chair*), Former U.S. Senator for New Mexico
STUART ANDREASON, Director, Center for Workforce and Economic Opportunity, Federal Reserve Bank of Atlanta
JOSH CARPENTER, Director, Economic Development, City of Birmingham, Alabama
KAREN ELZEY, Associate Executive Director, Workcred
AL FITZPAYNE, Executive Director, Future of Work Initiative, Aspen Institute
MARIA FLYNN, President and Chief Executive Officer, Jobs for the Future
VAN FREEMAN, Director, Future Workforce Development and Strategy, Aerospace Industries Association
DONNA GINTHER, Roy A. Roberts Distinguished Professor of Economics and Director for the Institute for Policy and Social Research, University of Kansas
RACHEL LIPSON, Director, Project on Workforce, Harvard University
ELOY ORTIZ OAKLEY, Chancellor, California Community Colleges
LEE WELLINGTON, Executive Director, Urban Manufacturing Alliance

Project Staff

TOM RUDIN, Senior Board Director (until December 2020)
LIDA BENINSON, Senior Program Officer
INEZ ADAMS, Senior Program Assistant (until November 2020)
SELAM ARAIA, Senior Program Assistant
FRED LESTINA, Research Associate (until September 2020)

Consultant

JOE ALPER, Consulting Writer

---

[1] The National Academies of Sciences, Engineering, and Medicine's planning committees are solely responsible for organizing the workshop, identifying topics, and choosing speakers. The responsibility for the published Proceedings of a Workshop rests with the workshop rapporteurs and the institution.

BOARD ON HIGHER EDUCATION AND WORKFORCE[1]

KUMBLE R. SUBBASWAMY (*Chair*), University of Massachusetts, Amherst
ANGELA BYARS-WINSTON, University of Wisconsin-Madison
JAIME CURTIS-FISK, The Dow Chemical Company
MARIELENA DESANCTIS, Broward College
APRILLE J. ERICSSON, National Aeronautics and Space Administration Goddard Space Flight Center
JOAN FERRINI-MUNDY, University of Maine
GABRIELA GONZALEZ (NAS), Louisiana State University
TASHA R. INNISS, Spelman College
SALLY K. MASON, University of Iowa
DOUGLAS S. MASSEY (NAS), Princeton University
RICHARD K. MILLER (NAE), Olin College of Engineering
PATRICIA SILVEYRA, University of North Carolina at Chapel Hill
KATE STOLL, Massachusetts Institute of Technology's Washington Office
MEGHNA TARE, University of Texas
MARY WOOLLEY (NAM), Research! America

[1] The National Academies of Sciences, Engineering, and Medicine's boards, forums, and roundtables do not issue, review, or approve individual documents. The responsibility for the published proceedings of a virtual workshop series rests with the workshop rapporteurs and the institution.

Board on Higher Education and Workforce Staff

LEIGH MILES JACKSON, Acting Director
THOMAS RUDIN, Director (until November 2020)
AUSTEN APPLEGATE, Research Associate
ARIELLE BAKER, Associate Program Officer
ASHLEY BEAR, Senior Program Officer
LIDA BENINSON, Senior Program Officer
FRAZIER BENYA, Senior Program Officer
IMANI BRAXTON-ALLEN, Senior Program Assistant (until April 2021)
MARIA LUND DAHLBERG, Senior Program Officer
ALEX HELMAN, Program Officer
REBEKAH HUTTON, Program Officer
PRIYANKA NALAMADA, Associate Program Officer
LAYNE A. SCHERER, Senior Program Officer
JOHN VERAS, Senior Program Assistant
MARQUITA WHITING, Senior Program Assistant

Acknowledgment of Reviewers

This proceedings of a virtual workshop series was reviewed in draft form by individuals chosen for their diverse perspectives and technical expertise. The purpose of this independent review is to provide candid and critical comments that will assist the National Academies of Sciences, Engineering, and Medicine in making each published proceedings as sound as possible and to ensure that it meets the institutional standards for quality, objectivity, evidence, and responsiveness to the charge. The review comments and draft manuscript remain confidential to protect the integrity of the process.

We thank the following individuals for their review of this proceedings: LaChaun Banks, Harvard University; Maria Cormier, Columbia University; Anna Duran, Avatar Research Institute; and Alexandra Beatty, the National Academies of Sciences, Engineering, and Medicine.

Although the reviewers listed above provided many constructive comments and suggestions, they were not asked to endorse the content of the proceedings nor did they see the final draft before its release. The review of this proceedings was overseen by George Boggs, Palomar College. He was responsible for making certain that an independent examination of this proceedings was carried out in accordance with standards of the National Academies and that all review comments were carefully considered. Responsibility for the final content rests entirely with the rapporteurs and the National Academies.

Acknowledgments

The National Academies of Sciences, Engineering, and Medicine's Board on Higher Education and Workforce thanks the Lumina Foundation for its support of this workshop series and wishes to express its sincere gratitude to the planning committee chair Jeff Bingaman for his valuable contribution to the development and orchestration of this virtual workshop series. We also wish to thank all the members of the planning committee, who collaborated to ensure a workshop complete with informative presentations and rich discussions. Finally, we want to thank the speakers and moderators, who generously shared their expertise and their time with workshop participants.

Support from the many annual sponsors of the Board of Higher Education and Workforce is critical to the board's work. The sponsors include the Department of Health and Human Services, National Endowment for the Arts, National Endowment for the Humanities, National Institutes of Health, National Science Foundation, Institute for Education Sciences, Alfred P. Sloan Foundation, Andrew W. Mellon Foundation, Burroughs Wellcome Fund, Elsevier Corporation, Helmsley Charitable Trust, Howard Hughes Medical Institute, Lumina Foundation, National Math and Science Initiative, and Spencer Foundation.

Contents

1 INTRODUCTION 1

2 BIRMINGHAM, ALABAMA 5

Understanding the Economic and Workforce Impacts of COVID-19 on Birmingham, 6

Birmingham's Economic Outlook Prior to and During COVID-19, 12

Changes in Workforce Training Programs in Birmingham, 16

Higher Education Programming for the Birmingham Workforce, 21

Closing Remarks, 26

3 BOSTON, MASSACHUSETTS 29

Understanding the Economic and Workforce Impacts of COVID-19 on Boston, 31

A Recovery with Equity, 35

COVID Response and Workforce Transitions in Greater Boston, 36

COVID-19 and Boston's Economic Divide: The Intersection of the Pandemic, Racial and Socioeconomic Inequality, and Access to Good Jobs in STEMM, 43

Catalyzing STEMM Entrepreneurship for Economic Mobility, 48

Closing Remarks, 53

4 RICHMOND, VIRGINIA 57
Understanding the Economic and Workforce Impacts of COVID-19 on Richmond, 57
Response of Richmond's Education and Training Infrastructure to the COVID-19 Crisis, 64
Innovative Workforce Solutions: How State and Local Policy and Programs Are Meeting the COVID-19 Challenge, 67
Closing Remarks, 71

5 RIVERSIDE, CALIFORNIA 73
Understanding the Economic and Workforce Impacts of COVID-19 on Riverside, 73
Greater Riverside Economic Outlook Prior to and During COVID-19, 77
Regional Ecosystems and Opportunities, 82
Higher Education's Role in Economic Mobility, 85
Closing Remarks, 89

6 WICHITA, KANSAS 91
Understanding the Economic and Workforce Impacts of COVID-19 on Wichita, 91
State and Local Policy Responses, 95
Response of Wichita's Higher Education Workforce Training Efforts to the COVID-19 Crisis, 99
Response of Wichita's Workforce Training Infrastructure to the COVID-19 Crisis, 102
Closing Remarks, 105

7 REGIONAL OBSERVATIONS WITH NATIONAL IMPLICATIONS 107
Regional STEMM Workforce Training Responses in the Wake of COVID-19, 108
Digital Infrastructure to Equitably Support Workforce Training, 114
Key Services That Support Equitable Access to Workforce Training, 115
Future Design of Workforce Training, 121
Common Themes Discussed Across the Workshop Series, 125

REFERENCES 129

APPENDIXES

A Statement of Task 131

B Workshop Agendas 133

C Planning Committee Biographies 149

Box and Figures

BOX

7-1 Common Themes Across the Six Workshops, 126

FIGURES

2-1 Demographics of the city of Birmingham and the Birmingham metropolitan area, 7
2-2 Educational attainment in the city of Birmingham and the Birmingham metropolitan area, 8
2-3 Unemployment claims for Alabama by age and race, 9
2-4 Unemployment claims for Alabama by industry, 10
2-5 The impact of Innovate Birmingham, 20

3-1 Demographics of the greater Boston metropolitan area and the city of Boston, 31
3-2 Educational attainment in the city of Boston and the greater Boston metropolitan area, 32
3-3 Unemployment claims for Massachusetts by age and race, 33
3-4 Unemployment claims for Massachusetts by industry, 34
3-5 Skill shapes for southern and northern California illustrating the different skill sets needed for regional manufacturing industries, 50
3-6 Skill shapes for digital marketing roles in Denver, Atlanta, and Boise, 51

4-1 Demographics of the city of Richmond and the Richmond metropolitan area, 59
4-2 Educational attainment in the city of Richmond and the Richmond metropolitan area, 60
4-3 Unemployment claims for Virginia by age and race, 61
4-4 Unemployment claims for Virginia by industry, 62

5-1 Demographics of the city of Riverside and the Riverside metropolitan area, 74
5-2 Educational attainment in the city of Riverside and the Riverside metropolitan area, 75
5-3 Unemployment claims for California by age and race, 76
5-4 Unemployment claims for California by industry, 77

6-1 Demographics of the city of Wichita and the Wichita metropolitan area, 92
6-2 Educational attainment in the city of Wichita and the Wichita metropolitan area, 93
6-3 Unemployment claims for Kansas by industry, 94
6-4 Unemployment claims for Kansas by age and race, 95

1

Introduction

The COVID-19 pandemic is transforming the global economy and significantly shifting workforce demand, requiring quick, adaptive responses. The pandemic has revealed the vulnerabilities of many organizations and regional economies, and it has accelerated trends that could lead to significant improvements in productivity, performance, and resilience, which will enable organizations and regions to thrive in the "next normal." What is unclear is how demands in the science, technology, engineering, mathematics, and medicine (STEMM) workforce have changed and whether there is an opportunity to upskill, retrain, and certify a significant portion of displaced workers to meet new and shifting labor demands in a way that is equitable, inclusive, and resilient to global instability.

To explore how communities around the United States are addressing workforce issues laid bare by the COVID-19 pandemic and how they are taking advantage of local opportunities to expand their STEMM workforces to position them for success going forward, the Board of Higher Education and Workforce of the National Academies of Sciences, Engineering, and Medicine (the National Academies), with financial support from the Lumina Foundation, established an ad hoc committee to plan a series of workshops to identify immediate and near-term regional STEMM workforce needs in the wake of the COVID-19 pandemic. The complete Statement of Task for this project is included in Appendix A. The planning committee identified five U.S. cities and their associated metropolitan areas—Birmingham, Alabama; Boston, Massachusetts; Richmond,

Virginia; Riverside, California; and Wichita, Kansas—to host workshops highlighting promising practices that communities can use to respond urgently and appropriately to their STEMM workforce needs. The planning committee also organized a sixth workshop to discuss how the lessons learned during the five region-focused workshops could be applied in other communities to meet STEMM workforce needs.

The committee considered a variety of factors in selecting cities for the workshop series. It aimed to select cities that differed from one another by geographic region, population size and density, racial and ethnic demographics, educational attainment, and employment trends affected by COVID-19. The committee intended for the cities examined in the workshop series to cover a broad range of attributes so that cities not examined in the series could find some applicable or overlapping scenarios for meeting their own regional STEMM workforce needs. Committee members' networks for engaging local stakeholders and identifying speakers were integral to the workshop planning efforts as well. Each workshop began with an overview of the selected metropolitan areas' employment and education situation that included data from the Federal Reserve and other sources. The committee designed the workshop series so that workforce and education data were arranged for each region in a consistent graphical format for comparison purposes.

The planning committee designed the workshops to highlight promising practices that community colleges and baccalaureate-granting institutions of higher education, as well as non-academic training and certification programs, can use to respond urgently and appropriately to those needs. The speakers, panelists, and workshop participants presented a broad range of views and ideas that aimed to address the following questions in the Statement of Task:

- How have the current and near-term regional STEMM labor market needs changed as a result of the COVID-19 pandemic, particularly the needs for the skilled technical workforce, such as information technology and computing fields, health technology sectors, and manufacturing?
- To what extent can regions create a profile that links STEMM educational and training resources with COVID-19–impacted workforce needs, including retraining and certification of professionals?
- What practices and policies can local higher education institutions and workforce training organizations adopt to respond to local

industry STEMM workforce needs, both individually and as part of regional coalitions? Which effective practices and policies are replicable and scalable? What barriers, if any, exist that inhibit these programs from meeting COVID-19 shifts in regional STEMM workforce needs?

- How can workforce training and education systems address the significant and urgent needs of the underrepresented minority population, which has been especially impacted by the effects of COVID-19? Can the recent events related to the death of Black men and women at the hands of police and the Black Lives Matter movement raise awareness of the disenfranchisement of the nation's underrepresented minority populations that result from institutionalized racism, and therefore increase focus on the needs of those populations?
- What actions are needed to assist local higher education institutions, training and certification programs, industry, and policy makers at the local, state, and federal levels in fostering improved linkages between education and training resources and STEMM workforce needs at the regional level as a result of the COVID-19 pandemic?

This proceedings of a virtual workshop series summarizes the presentations and discussions from the six public workshops that made up the virtual workshop series and highlights the key points raised during the presentations, moderated panel discussions and deliberations, and open discussions among the workshop participants. In accordance with the policies of the National Academies, this proceedings has been prepared by the workshop rapporteurs as a factual summary of what occurred at the workshops. The planning committee's role was limited to planning the workshop. Statements, recommendations, and opinions expressed are those of individual presenters and participants and are not necessarily endorsed or verified by the National Academies of Sciences, Engineering, and Medicine, and they should not be construed as reflecting any group consensus.

The workshops did not attempt to establish any conclusions or recommendations about needs and future directions, focusing instead on issues identified by the speakers and workshop participants. During a planning meeting preceding the workshop series, the committee selected a variety of themes to discuss throughout the workshops relevant to addressing STEMM workforce shifts resulting from the COVID-19 pandemic. The themes the committee selected included job training offerings; valuing, managing, and leveraging diversity and equity principles and practices as key components of workforce training; degrees and credentials; the digital divide; social support services; and data. Box 7-1 includes additional details on these themes.

2

Birmingham, Alabama

Birmingham, Alabama, like the rest of the nation, was facing a variety of workforce challenges prior to March 2020, but the COVID-19 pandemic raised new challenges and perhaps unique opportunities as well. The committee selected Birmingham because it is a city in the geographic south of the United States with stark demographic and employment differences between the city and surrounding region. Josh Carpenter, director of economic development for the city of Birmingham, opened the workshop by sharing the motivation to discuss how the changes brought about by the pandemic "can be beneficial for those seeking new jobs or career paths here in Birmingham." He noted that the pandemic is not the only factor influencing the nature of work in Birmingham. The Black Lives Matter movement in particular is shining a light on persistent inequities embedded in the workforce culture. "It is clear that efforts to rebuild our workforce must include intentional actions to root out racism and create inclusive conditions to support diversity and equity in all job sectors," said Carpenter. "As we continue these conversations, we want to make sure that piece of this struggle is front and center."

This workshop began with sessions that presented data on the economic outlook, labor force, educational attainment, and demographic data for the Birmingham metropolitan area, and discussed Birmingham's economic outlook prior to and during the COVID-19 pandemic and its investments in workforce training programs. These were followed by panel discussions on emerging workforce training programs, efforts to address labor inequities exacerbated by the pandemic, and efforts by institutions of higher education in the region to support local workforce development efforts.

UNDERSTANDING THE ECONOMIC AND WORKFORCE IMPACTS OF COVID-19 ON BIRMINGHAM

Prior to the COVID-19 pandemic, the city of Birmingham—the fourth-largest majority Black city in America—had undergone a rigorous study using quantitative and qualitative methods to better understand its workforce needs. Carpenter noted that city leaders and activists had been thinking hard not just about the ways in which equity should be morally important, but also how it is vital to the city's economic viability and competitiveness.

To provide context for the ensuing discussion, Stuart Andreason, director of the Center for Workforce and Economic Opportunity at the Federal Reserve Bank of Atlanta, provided a data-based overview of Birmingham's economic situation, noting that some of Birmingham's unique characteristics point to regional variations that affect the city's eventual recovery from the pandemic-induced recession and foreseeing this as a theme throughout the workshop series. He also commented that a challenge for anyone who uses data to think about economic issues and to inform policy and practice is that most data are not current. In fact, he said, there exist few pandemic-era labor and economic data at this time.

Demographics and Educational Attainment

Andreason said that the city of Birmingham entered the COVID-19 pandemic with greater economic stresses than the city's suburbs were experiencing. For example, the poverty rate in the city proper was 25.2 percent versus 14 percent for the larger Birmingham metropolitan area, and median wages for people 16 years and older was $30,014 in the city compared to $36,427 for the metro area.[1] He noted that while Birmingham city is majority Black, the metro area is majority white (Figure 2-1). A comparison of educational attainment shows less variation between the city and metro area than one might expect, he added. Regarding the highest level of educational attainment in the metro area, roughly half the population

---

[1] These data are from the 2019 American Community Survey single-year estimates, available at https://www.census.gov/newsroom/press-kits/2020/acs-1year.html.

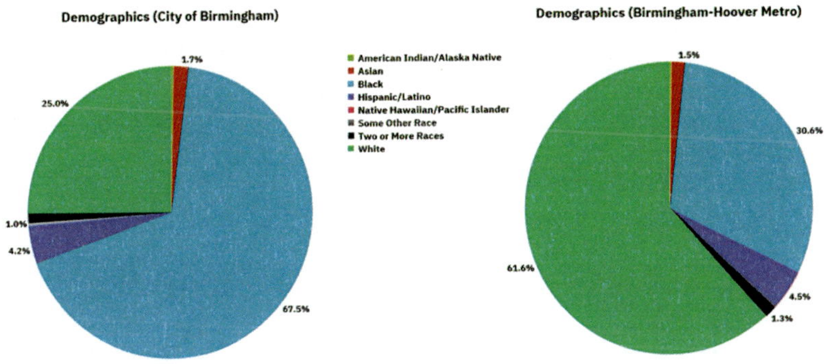

FIGURE 2-1 Demographics of the city of Birmingham (left) and the Birmingham metropolitan area (right).
SOURCE: Andreason slides 3 and 4, based on data from the 2019 American Community Survey single-year estimates, are available at https://www.census.gov/newsroom/press-kits/2020/acs-1year.html.

has less than an associate degree, more than 25 percent attained a high school diploma, and about 10 to 12 percent of the population did not complete high school (Figure 2-2). The only significant difference for the city is that a slightly higher proportion of its population did not get a high school diploma. "We know that workers that have a high school degree or less are significantly challenged and that many of the job losses have been concentrated in lower-skilled, lower-educational-expectation jobs in the crisis," said Andreason.

Unemployment Data

Metro Birmingham's unemployment rate in 2019 was below 4 percent, which put it among the nation's metropolitan areas with the lowest unemployment rates. Unemployment in the city, however, was considerably higher in 2019, at 7.3 percent. In 2020, unemployment in the metro area spiked in March and April, as it did throughout the country, and while it has declined since then, unemployment remains around three times higher than in the same period in 2019.[2] Unemployment claims data for the state of Alabama—data unavailable for Birmingham city or metro area—show that

[2] Comparable data for the city of Birmingham were not yet available at the time of this workshop.

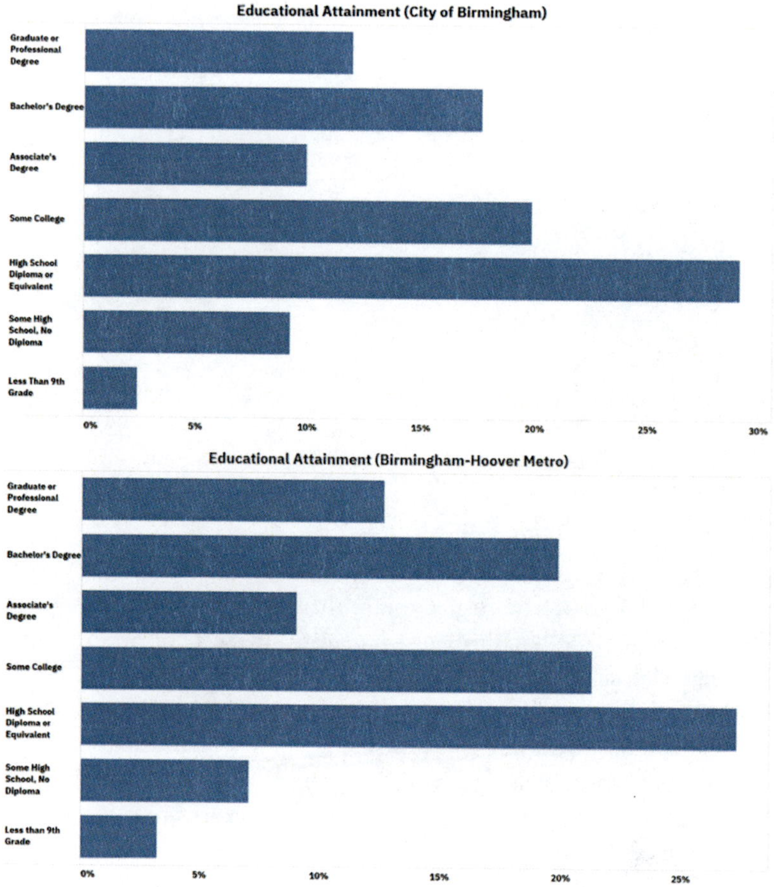

FIGURE 2-2 Educational attainment in the city of Birmingham (top) and the Birmingham metropolitan area (bottom).
SOURCE: Andreason slides 5 and 6, based on the 2019 American Community Survey single-year estimates, are available at https://www.census.gov/newsroom/press-kits/2020/acs-1year.html.

women have fared significantly worse than men during the pandemic. In September 2020, 57 percent of the claimants in Alabama were women, even though women make up only 47 percent of the state's workforce. Similarly, 46 percent of the claimants were under age 35 and 53.5 percent were Black, even though those two demographic groups represent only 35.8 percent and 24.9 percent, respectively, of the state's workforce (Figure 2-3). Andreason noted that during the week of September 19, more workers were on the new

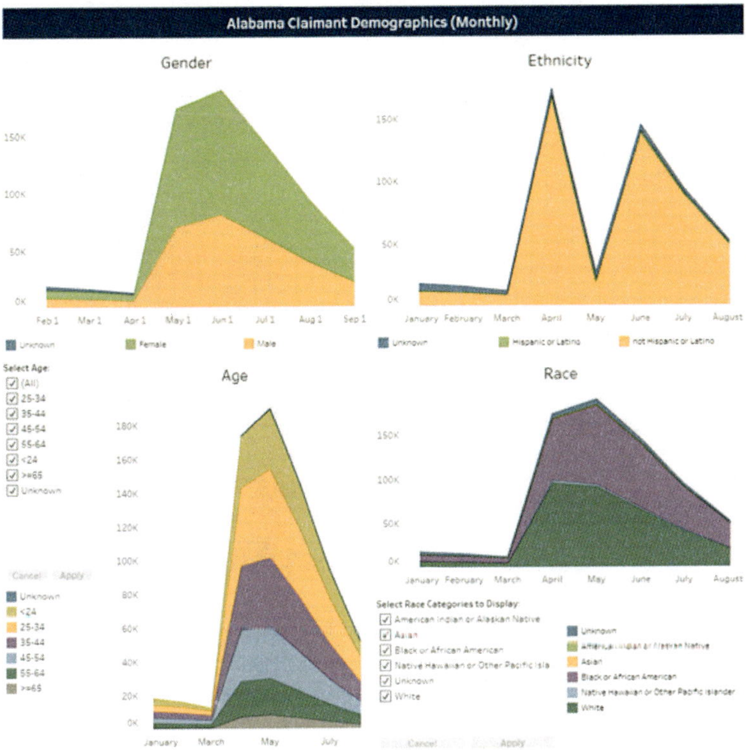

FIGURE 2-3 Unemployment claims for Alabama by age and race.
SOURCE: Andreason slide 8.

Pandemic Unemployment Assistance Program for nontraditional workers, part of the Coronavirus Aid, Relief, and Economic Security (CARES) Act, than were on traditional unemployment. Carpenter commented that a disproportionate percentage of Black unemployment speaks both to job quality and to who has access to jobs that pay a family-supporting wage.

Looking at the weekly employment claims by industry across Alabama showed that 15 percent of the claims in September 2020 were from workers in the food industry; 13 percent were in administrate support and waste management, which includes staffing firms; 10.5 percent were in manufacturing; 9.5 percent were in retail trade; 7.3 percent were in healthcare, and 3.9 percent were in construction (Figure 2-4).[3] Andreason pointed out that

[3] Unemployment claims by industry were only available by state, and not by city or metropolitan area.

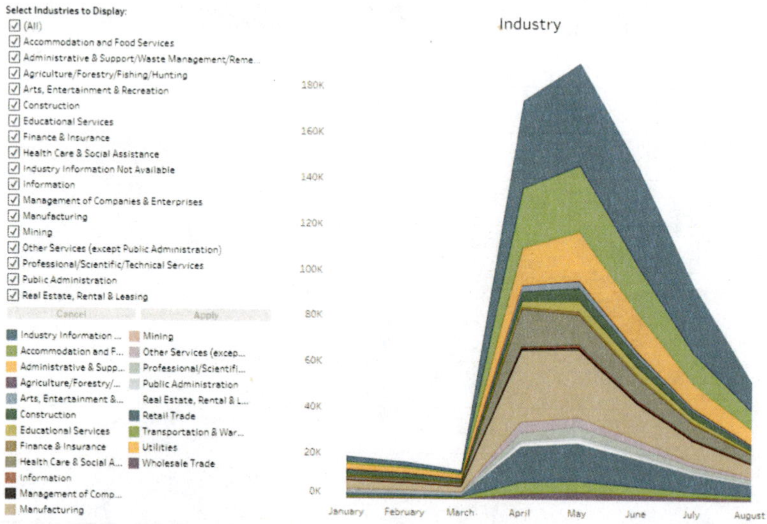

FIGURE 2-4 Unemployment claims for Alabama by industry.
SOURCE: Andreason slide 9.

a significant portion of the claims in Alabama were in jobs that are difficult to do remotely. In terms of the risk of permanent job loss arising from the pandemic, he explained that the Atlanta Federal Reserve's analysis suggested that the total risk of employment loss in Birmingham was lower than for other southern cities because of its mix of employers. The Atlanta Federal Reserve's analysis did show, however, that Birmingham was at higher risk of permanently losing what economists call middle-skill jobs—those that pay above the median wage—compared to other southern cities. He noted that looking just at so-called opportunity jobs—those that do not require a bachelor's degree but pay over the median wage—shows that many at risk of being lost permanently are those that are difficult to perform remotely, such as repair workers, insurance claim processors, and carpenters. What these data point to, said Andreason, is the critical need to focus on helping people who were in high-quality jobs likely to be lost permanently to transition to future opportunities.

Metro Birmingham's industry mix, he continued, suggests that it might be somewhat more insulated from the risk of permanent job loss. Estimates from the Congressional Budget Office in 2006 on how a potential influenza pandemic would affect regional economies suggested that Birmingham's metro area would be slightly less affected than would Alabama as a whole, though the effect would still be large (Arnold et al., 2006). Pre-pandemic estimates suggest that 28.6 percent of Birmingham's workers would be able to work from home, which Andreason said was comparable to the state estimate as well as those for many metropolitan areas nationwide (BLS, 2019a, 2019b).

Internet Connectivity

The ability to work from home during the pandemic has had a strong impact on individuals' possibility of keeping a job that converted to a remote environment or gaining employment that requires remote work, as well as adults' and children's ability to do remote learning and training. In this respect, broadband internet access, which gives a person access to the internet on a useful computing device for work and learning, is a critical consideration. An analysis of broadband access by census track for the Birmingham metro area suggests that internet access is a regionwide issue affecting residents in urban, exurban, and rural areas alike, though the reasons why a given household may not have broadband internet can include not having it available in their neighborhood, not being able to afford it, and not choosing to subscribe to it. Nearly 85 percent of all households in the Birmingham metro area have a high-speed internet subscription, compared to 76.9 percent of households in the city proper. Almost 17 percent of households in the metro area access the internet only through a cellular data plan, compared to 20.6 percent in the city. Just over 40 percent of households in both the metro area and the city earning less than $20,000 have no internet subscription at all. Across the metro area, 72.8 percent of all households have a desktop or laptop computer, compared to 60.6 percent in the city; 12.7 percent in the metro area and 19.1 have only a smartphone; and 8.5 percent in the metro area and 13.8 percent in the city have no computing device through which to access the internet. "About one in five households in the city of Birmingham access the internet on their cellphone only," said Andreason. "Imagine doing work, school, or job training on a cellphone only." In other words, he said, internet access is important, but so, too is the hardware people use to interface with the internet when thinking about developing learning and work platforms and designing interventions to help people navigate the current crisis. He noted that these challenges can have a disproportionate impact on women and Black residents.

BIRMINGHAM'S ECONOMIC OUTLOOK PRIOR TO AND DURING COVID-19

The workshop's first panel session featured three speakers who discussed pre-COVID investments in workforce training programs and the impact of the pandemic on their workforce alignment and learner access. The speakers were Joseph Parilla, metropolitan policy program fellow of the Brookings Institution, and Sanjay Singh, advisor and board member of Pack Health, which has developed a patient-centered digital health coaching program for individuals with chronic health issues. After the two short presentations, Antiqua Cleggett, executive director of Central Six AlabamaWorks!, the state's workforce and economic development agency serving the Birmingham metropolitan area, moderated a discussion among the panelists and workshop participants.

Speaking first, Parilla noted that the Brookings Institution has partnered with the city of Birmingham and a network of corporate and university leaders since 2019 to develop an inclusive economic development strategy for the following reasons:

- Birmingham, as a majority Black city in an older industrial region, is a bellwether for many of the challenges the country has been facing regarding structural racism.
- There was a change moment with the election of a new mayor and a collection of corporate and civic leaders who recognized that it was time to act given that the interests and goals of the city's public, private, and civic sectors were aligned.
- Birmingham has significant assets upon which to build an economic development strategy, identify labor market opportunities, and accomplish something significant for the city.

Parilla explained that a new mechanism called Prosper Birmingham, established by a group of local leaders and funders, was used to carry out a market assessment of the economic conditions of the region and identified key factors affecting the situation. An important source of information for this effort was a labor market assessment from the Building (it) Together coalition (Bradley et al., 2018), and, along with an alliance with the Brookings Institution, they identified disconnects between education and training efforts and the job requirements of local workplaces seeking to fill positions. "That foundation led to a new understanding of where there are in-demand career opportunities and how that aligns with the education system," said Parilla.

Another critical piece of this effort was the availability of a new generation of in-demand training programs designed to complement traditional training programs in a way that recognizes the challenges of working with people who may come from a background of economic scarcity. These programs provide wrap-around services to stabilize the trainees' lives and allow them to upscale their skills into something that is in demand. As an example, he cited Innovate Birmingham, which provides several programs to meet the demand for technology talent among employers in the city without requiring people to go the route of attending a two- or four-year institution. These programs, said Parilla, offer curricula informed by industry and provide necessary resources to help those who enroll in the program to address barriers in their lives. Another of the city's programs is the Birmingham Promise, which Parilla called a generational investment in the city's young people. Birmingham Promise offers both college scholarships for students in Birmingham's city schools and apprenticeship opportunities at local companies.

These types of programs, said Parilla, tell participants that if they take advantage of educational and training opportunities, the city will have their back and invest in them. "That is an incredible civic gesture that the mayor has led, and that the corporate community has rallied behind, but it is also smart because the future of the Birmingham economy, those employers, and any sort of successful civil society is those kids," he said. In closing, Parilla said that for all of Birmingham's challenges that Andreason laid out, leadership across the city is behind these programs, which is why he and his colleagues at Brookings are "super bullish" on Birmingham's future.

Fulfilling a Need for Health Advisors

Singh opened his remarks by noting that Birmingham has undergone a massive transition since 1993, having worked through four iterations of change and now starting its fifth. He then explained that Pack Health is a digital health advising company that helps people with chronic conditions manage their daily health when they are not interfacing with a healthcare system. One of Pack Health's tasks is to develop a workforce of health advisors, a position for which no college or university in the country offers a degree. Instead, the company recruits community health workers, registered dieticians, social workers, nutritionists, and other people from an allied health background and puts them through an intense, six-month training program that it designed and that has now become virtual. Over time, each health advisor will specialize in a platform area or condition, such as oncology or cardiometabolic disorders. Singh noted that all of the health advisors' interactions with their clients are virtual, taking place through either email, phone calls, or texts. He explained that every client is assigned to a single health advisor for at least 90 days and that the health advisors are only available during set times arranged between the client and advisor, not 24 hours a day, seven days a week.

As this field of health advisor has evolved and enough people have taken this career path, the National Board of Medical Examiners created a curriculum for certification as a health and wellness coach.[4] Pack Health has adopted this certification, and, in addition to coursework, requires all its students to coach for 1,000 hours before becoming eligible to take the certification exam. Singh noted that since this program is not the company's primary source of income, it has set the price of enrollment to be the most affordable in the job training marketplace, allowing displaced workers to make a career change or sharpen their skills. In addition, his company is working with Carpenter's Birmingham Strong initiative to provide a pathway for displaced employees who do not have health backgrounds to become customer success representatives. This position serves as an assistant to the health advisors, in much the same way that a pharmacy technician assists a pharmacist. With experience and ongoing interest, customer success representatives can apply for the health advisor program. For those individuals, the company pays their tuition, fees, and books. As a final comment, Singh explained that since the company was created to offer virtual programs, there has been minimal disruption to its programs during the COVID-19 pandemic. In fact, he added, the company has tripled in size since March 15, 2020, and expects to double again over the next 12 months.

Discussion

The biggest changes Parilla has observed in Birmingham's workforce training programs since COVID's emergence have been educational and training institutions learning how to operate in a remote, online environment and the creation of a crisis mitigation strategy called Birmingham Strong. Birmingham Strong's mission is to take displaced workers and serve

[4] Additional information is available at https://nbhwc.org/.

as the bridging mechanism between idle workers and efforts to address the most significant social and public health problems in the city triggered by the pandemic. While Birmingham Strong realized quickly that it would not be able to place every unemployed worker, it could steer a significant number of them into this program and pay them $15 to $25 an hour. Parilla's colleagues at the Brookings Institution wrote about this program and now other cities are trying the same approach (Ross et al., 2020). He noted, too, that future federal relief packages might include efforts to scale this program, given the long-term effects of the pandemic that will leave some workers unemployed for a considerable time and create civic and social challenges. Singh added that the pandemic is forcing organizations that need employees to hire and train them remotely, which few are set up to do.

Regarding potential barriers to workforce development partnerships, Parilla suggested identifying sectors in the local environment that have a big enough need for employees that employers will be willing to put the time and resources into the partnership. He said that resources will always be a challenge, for even though moving someone from a training program into a career pathway is less expensive than sending them to a four-year degree program, doing so still costs $8,000 to $10,000, which may not be affordable for someone who is unemployed. Equally important, said Cleggett, is helping displaced workers pull together their workforce profiles that can help them understand where they have transferrable skills and to provide help navigating the training and employment systems, which are not always user-friendly.

To assist higher education institutions, industry, and state and federal policy makers in meeting workforce demands resulting from the COVID-19 pandemic, Parilla suggested operationalizing remote training and using the pandemic to further higher education's use of online learning. Such actions would facilitate lifelong learning and help these institutions sustain themselves into the future. He called on the federal government to use this historic labor market crisis as an opportunity to invest 10 times as much as usual in workforce development so that institutions can reach people on the scale needed to address the sheer number of displaced workers. Singh noted the need for more micro-credentialing opportunities that students could earn while still in school to short-circuit the need to further train an individual after they have received their college degree. The challenge, he said, is to compress the time scale required for new employees to become productive employees.

CHANGES IN WORKFORCE TRAINING PROGRAMS IN BIRMINGHAM

In the workshop's second session, four panelists discussed emerging workforce training programs for the Birmingham region and efforts to address labor inequities exacerbated by the pandemic. The panelists were Suzanna Fritzberg, executive director of Birmingham Strong; Rachel Harmon, chief executive officer of Birmingham Promise, a program that provides college scholarships and apprenticeship opportunities at local companies for Birmingham high school students; Waymond Jackson, chief executive officer of Ed Farm, an organization that promotes innovation in education to increase educational equity, improve learning outcomes through technology, and prepare the future workforce; and Emily Wykle, acting principal investigator and board member of Innovate Birmingham—a coalition of more than 15 community partners and 30 employers working to meet the rapidly growing local demand for an information technology (IT) workforce by harnessing the potential of under- and unemployed residents in Central Alabama—and director of external affairs at the University of Alabama at Birmingham. LaChaun Banks, associate director of the Ash Center's Innovations in Government Program at the Harvard Kennedy School, moderated a discussion among the panelists and workshop attendees.

Birmingham Strong

Birmingham Strong's flagship program is the Birmingham Service Corps, launched in May 2020, and its major funder is the city of Birmingham, though all its activities are designed to be reimbursable through the CARES Act. Fritzberg noted that Birmingham residents were already experiencing extreme levels of poverty—nearly half of the city's families lived in poverty before the pandemic—and many neighborhoods in the city are no stranger to some of the dire unemployment numbers seen now, particularly neighborhoods that are majority Black.

Birmingham Strong modeled the Service Corps on some of the Great Depression–era work relief programs with the goal of connecting unemployed residents with paid volunteer opportunities that address community needs. In general, the program prioritizes building skills among the participants for growth industries that will best position the participants to compete for jobs in the future. At the time of the workshop, the program had placed over 350 residents into paid work positions that pay between $16 and $22 an hour. Two-thirds of corps members placed into paid positions are people of color, and 63 percent are Black. Fritzberg said that for 90 percent of the participants, Service Corp work is a key factor in their ability to pay their rent or utility payments, not fall behind on debt repayment, and meet basic economic needs. In addition, over 2,800 small businesses have been assisted through a call center, staffed by Corps members, that provides CARES federal assistance advice and administers small business technical assistance programs.

According to the *Building (it) Together* report, new job postings in Birmingham are down by nearly 50 percent, and the Birmingham metropolitan area stands to lose 14,000 jobs for individuals with a high school diploma or less by the fourth quarter of 2024 (Bradley et al., 2018). Fritzberg said that the skills acquired in the Birmingham Service Corps can be the foundation for new career trajectories in data collection and management, client services, and precision population health, an emerging industry. Toward that end, the Birmingham Strong model leverages COVID relief and recovery funding toward earn-and-learn workforce development rooted in employer partnerships and targeted toward community need. The program focuses on workers between ages 32 and 55 and could evolve in ways that more closely align it with the formal workforce ecosystem and a regional workforce development strategy.

Birmingham Promise

Harmon described how Birmingham Promise started as a campaign promise to capitalize on the inherent value of local talent to strengthen the city's economy, eventually became one of the city's strategic initiatives, and now functions as an independent nonprofit organization. The organization has two programmatic streams: a scholarship program similar to the more than 300 other Promise programs across the country, and a youth apprenticeship program. The scholarship program focuses on providing postsecondary school access and ensuring postsecondary success for students in the city's schools, over 90 percent of whom are Black and about half of whom come from families that experience poverty.

In the apprenticeship program, juniors and seniors in high school participate in paid, work-based learning opportunities. The program focuses on four industry clusters based on the research done for the *Building (it) Together* report (Bradley et al., 2018). "We wanted to ensure that we were prioritizing industries and occupations that are high opportunity and where students would have a higher chance of ending up in a job that is family-sustaining or a career that's family-sustaining," Harmon explained.

She noted that one of the most important elements of Birmingham Promise is that it is inherently collaborative and is done in close partnership with the city's employer community, its K-12 system, and the region's higher education institutions, including the University of Alabama at Birmingham. This multi-institutional partnership creates smoother pathways that are easier for students from underrepresented groups to navigate, which is important from a systems change perspective.

The first class of postsecondary scholarship students started in college in the autumn of 2020, and the second cohort of youth apprentices started in January 2021. Harmon noted that 75 percent of the program's students who started college in the fall said their college experience had been affected negatively by the pandemic, whether due to a parent losing a job or the challenge of attending school in an unfamiliar virtual learning environment.

Ed Farm Programs

Jackson applauded the city of Birmingham for using data to drive progress in programs, particularly as they relate to serving people in the community. One thing these data illustrated was that the city did not have many residents with the digital skills needed to thrive in the 21st century job market, or residents with the skills to meet the projected six-fold increase in computer science jobs that would be available over the next 10 years and would require a bachelor's degree. This realization mobilized Jackson and several colleagues to approach Apple and ask them to help make Birmingham the epicenter for education, technology, training, and coding. The result was Ed Farm, launched in February 2020 with the goal of providing all people—and he stressed *all*—with the tools they would need to fill and create the jobs of the future.

Those jobs of the future, said Jackson, require individuals who have a deep ability to think critically and use digital skills, including coding, to solve problems. To give people those skills, Ed Farm offers a program called Pathways, an adult learner boot camp that teaches Apple's relatively simple-to-learn coding language. This 11-week program, which meets for three hours two nights per week, provides city residents with the opportunity to become interested in coding and gain some initial skills that can perhaps interest them in computing careers or develop the type of transferrable skills that local employers have indicated they need.

Jackson explained that when the pandemic struck, it forced Ed Farm to shift to a virtual instructional model and provide Apple MacBooks and iPads to its Pathways participants. The benefits of going virtual, he noted, were that it sped up efforts to provide those computing devices and it eliminated the transportation barrier that some participants faced in terms of getting to class.

Innovate Birmingham

Innovate Birmingham was designed around two critical challenges facing the city, explained Wykle. The first was the fact that Birmingham is the second fastest growth market in the nation for IT jobs, and the local talent pool cannot keep up with that growth. The second challenge is that there are some 26,000 disconnected youth in the city who have great capability but have not had the "on ramps" to IT careers. As she put it, talent is distributed equally in society, but opportunity is not.

Innovate Birmingham started with a $6 million seed grant from the Department of Labor's America's Promise program and has grown into a workforce training program that provides multiple pathways to technology careers. Full-time, 13-week boot camps focused on software development and data analytics are the program's "bread and butter," said Wykle. Innovate Birmingham also provides Alabama's first registered IT apprenticeship program in collaboration with its workforce partner, Central Six. This is an earn-to-learn opportunity for people who want to get a foot in the working world and acquire experience in a low-risk manner, she explained.

At the time of the workshop, Innovate Birmingham had served 774 participants with a 92 percent retention rate to completion (Figure 2-5). Wykle noted that in addition to the program being free, program staff provide wrap-around services, such as childcare and housing assistance, as well as professional development support including coaching. The program's demographics mirror those of the city of Birmingham. Wykle also highlighted the fact that Innovate Birmingham has played a key role in shaping regional workforce development efforts in the IT realm along with its nonprofit and K-12 partners and in supporting students when they go to one of the regional institutions of higher education and then into the working world. She added that the program's technology council, comprising employer partners, higher education partners, and workforce

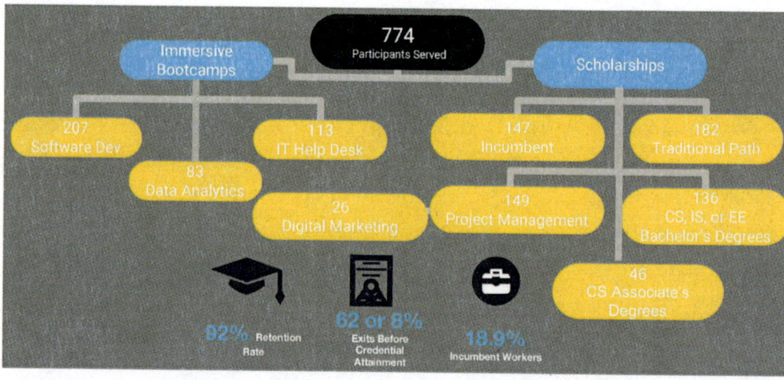

FIGURE 2-5 The impact of Innovate Birmingham.
SOURCE: Wykle slide 4.

partners, provides a venue to discuss what is needed to prepare students to be successful members of the IT workforce.

In terms of how the pandemic has affected Innovate Birmingham, Wykle said the biggest shift has been to move all of its boot camps and services to a virtual environment. This effort included creating a loaner program for computing devices such as laptops and tablets, and internet hotspots. In late spring 2020, there was a large increase in applications from furloughed and displaced workers, and the program has been able to serve many of these individuals, in part because the pandemic struck just as the program was already planning to expand across the six-county metropolitan region. She also credited several of the program's employer partners who designated scholarships targeted at people of color who lost jobs because of the pandemic to participate in Innovate Birmingham.

Discussion

When asked how Birmingham Strong's Service Corps was going to move forward once emergency federal funding to deal with the pandemic ends, Fritzberg said the program has some generous private sector philanthropic supporters in addition to its CARES Act funding. The hope, she said, is that by demonstrating that the Service Corps is having a real impact on the residents of Birmingham, this support will continue. In addition, the program receives a staffing fee for a significant portion of the jobs that the program supports. In response to a question about the types of skills Service Corps participants are gaining, some program participants are learning client management and customer support skills; others are learning data management skills. Currently, she said, the program does not yet have much data to identify the career paths that unfold from these experiences. The program is looking to expand its efforts on small business support by supporting Small Business Administration loan applications.

When asked how much interaction there is between Birmingham Strong and the traditional safety net to help people who are struggling financially to get into these programs, Fritzberg said that Birmingham Strong makes a point of employing individuals who are not eligible for unemployment. Jackson added that Ed Farm has let all of the local social services agencies and nonprofits know about the program's offering. He noted, too, that the corporate community is starting to provide computing devices and internet connections for city residents who are having to take advantage of safety net programs. Both Harmon and Wykle noted in closing that the pandemic has provided a way of focusing in a singular manner on addressing inequities, though it is still unclear whether these programs are helping to reduce inequities.

HIGHER EDUCATION PROGRAMMING FOR THE BIRMINGHAM WORKFORCE

The workshop's final panel session featured five speakers from institutions of higher education in Alabama: Jarralynne Agee, provost and senior vice president of academic affairs at Miles College; Cynthia Anthony, interim president of Lawson State Community College; Pam Benoit, senior vice president of academic affairs and provost at the University of Alabama at Birmingham; Guin Robinson, associate dean of economic development at Jefferson State Community College; and Daniel Coleman, president of Birmingham Southern College. Cassius Johnson, senior director of the Education Strategy Group, which works with education leaders and employers to design, scale, and implement strategies that improve attainment and advance equity, moderated the final discussion period.

Benoit noted that one consequence of the pandemic is that students are concerned about the types of jobs that will be available when they finish their degrees. She discussed the ongoing initiatives related to the science, technology, engineering, mathematics, and medicine workforce at the University of Alabama at Birmingham, including the development of new interdisciplinary bachelor's degree programs in bioinformatics, immunology, genetics and genomics sciences, cancer biology, and entrepreneurship. She also said that students have shown more interest in public health as a major since the start of the pandemic.

At the graduate level, the university is offering stackable certificates for specific competencies that can be turned into an interdisciplinary master of science. For example, a student who wants to work in research laboratory management might compile certificates in lab management and in leadership and professional development to create a master of science degree. To prepare to work in a nonprofit community health organization a student could combine a certificate in nonprofit management with one in public health. "There is a wide variety of different kinds of certificates that can be stacked together to be flexible and adaptive to changing workforce needs, especially in light of COVID," said Benoit.

Another initiative is the University of Alabama at Birmingham's Lifelong Learning program, led by a new associate provost charged with identifying opportunities to extend both credit and non-credit courses to respond to workforce development needs. Benoit said there is a great deal of interest in this program offering courses on issues such as mental health and resilience. The final initiative she discussed is the Alabama Advanced Specialized Skills, Education, and Training (ASSET) microcredentials program. The purpose of Alabama ASSET is to rapidly meet dynamic workforce needs for employment, advancement, or transferable skills for career mobility. Individuals earn these ASSET microcredentials by taking between six and 15 hours of graduate-level courses at any participating institution in Alabama. These credits can be applied later to a formal master of science program if desired.

Regarding workforce retention, the University of Alabama at Birmingham, the state's largest single employer, has established childcare programs for employees at its campus hotel that includes virtual learning assignments done with support of the university's students. The university has also partnered with the McWane Center—Birmingham's science museum—to provide full-day childcare and high-quality programming in science, technology, engineering, mathematics, and medicine during the COVID-19 pandemic.

Next, Robinson described Jefferson State Community College as an institution that serves a four-county area in central Alabama. Enrollment has dropped by just over 2 percent between fall 2019 and fall 2020 semesters. Jefferson State offers non-credit workforce education opportunities that provide skills training for specific, entry-level jobs. In this program, students acquire skills for immediate employment instead of credit toward a college degree. These nine-week to nine-month courses are offered across multiple sectors, including healthcare, IT, manufacturing, and industrial training, aligned with industry-recognized credentials. The courses are employer-led and informed by ever-changing labor market data. Jefferson State also offers several scholarships for healthcare training programs and for online IT programs directed toward women. It also has training programs for dislocated workers affected by COVID-19 layoffs that are funded by the Workforce Innovation and Opportunity Act. These programs include a line worker program with the Alabama Power Company using scholarships provided by an Appalachian Regional Commission grant, a vision assistant program in collaboration with the Callahan Eye Foundation, and a clinical medical assistant program in conjunction with the University of Alabama at Birmingham and St. Vincent's Hospital.

Miles College, said Agee, has changed its focus to one that aligns its academic programming with the workforce needs of Alabama's industries. The college's 100-plus-year history as a liberal arts institution means it can train its students on a wide range of subjects and have them be well versed in the information literacy, communication skills, and people skills they need to succeed in the business world. At the same time, said Agee, Miles College is helping its students build the technical skills to help them to be more successful at their jobs. Currently, given its partnerships with both local and national organizations, Miles College is focusing on artificial intelligence, security and cyber security, data science, and Swift coding. It is also launching an e-gaming program, both as an extracurricular e-gaming club and an interdisciplinary minor with business, communications, computer sciences, and psychology.

Agee explained that her institution is establishing internships, both in person and virtual, through the Birmingham Business Alliance and various federal agencies. Virtual internships are proving valuable for students with transportation issues or who cannot travel to Washington, DC, to participate in internships with federal agencies. Miles College is also creating certificate programs so that its students can build extra skills.

Lawson State Community College, explained Anthony, has chosen to use the pandemic as an opportunity for reinvention and to reimagine itself as an education and training entity. The pandemic has forced her and her colleagues to examine its processes and procedures and adopt new ways of doing things that include virtual, online, remote, hybrid, and distance learning in order to thrive both in the pandemic and post-pandemic era. Coming from a student services background, she stressed the importance of thinking about not just the teaching part of education but also how students learn and how they can embrace the training they receive.

Anthony also stressed the importance of collaborations and partnerships in her institution's retooling and redesigning activities so that it will have an impact on the workforce. She noted that in terms of programs, much of the activity is around expanding existing programs and adding hands-on training and simulations. For example, Lawson State Community College is part of a virtual reality–based workforce initiative sponsored by Alabama Power, the regional utility.

In the day's final panel presentation, Coleman provided a brief history of Birmingham Southern College, which was founded as a liberal arts college in 1856 as an institution that would serve the community and help develop a future workforce and leadership for the community. He then discussed data showing the top five "hard-to-fill" jobs in the Birmingham metropolitan area that require a bachelor's degree, both pre-COVID-19 and during the pandemic: registered nurse, sales representative, software developer/engineer, business development manager, and retail store manager. The highest-paying job on that list, according to Coleman, is software developer and engineer, which requires a broader skill set that grows broader every day.

Coleman acknowledged that the Birmingham metropolitan area's economy has been growing slowly for the past few decades, particularly compared to cities such as Nashville, Tennessee; Charlotte, North Carolina; and even Chattanooga, Tennessee. The key issue to him is that local employers are not filling these high-demand jobs with Birmingham residents, which in his experience in the business world and in talking with local companies means one thing: they are either hiring from other places or setting up offices outside of Birmingham.

Though Birmingham Southern is a small institution—its enrollment is approximately 1,100 students—it has produced many of the city's leaders, including 10 percent of the Birmingham bar and many leaders of local nonprofit organizations. However, going forward, he expects that the sectors of the local economy with leaders who graduated from Birmingham Southern will be much different than today. "Our challenge is to figure out how to keep this relevance and the importance of this institution here, and in doing so, we have to integrate new types of approaches, such as data science and technological skill sets, into the liberal arts curriculum, and demonstrate how we can apply these skills the same way that we apply other skills in liberal arts," said Coleman.

Discussion

Johnson observed that a common theme of the five presentations was the use of labor market information to inform the work the five higher education institutions are doing. Another theme, he said, was the ability or the need to be agile and pivot during the pandemic. However, several speakers noted challenges to engaging individuals who could benefit from further education and training and keeping them enrolled in the various programs their institutions offer. Anthony noted that Birmingham does a good job of connecting students to the wrap-around services they need to succeed at community colleges and four-year institutions. Getting the financial resources to create new programs, however, is another barrier. She also said that her institution, as well as the Alabama community college system overall, has put a great deal more focus on the enrollment management process and the student experience, from the initial recruitment through graduation and into the workforce. She added that it is doubly important now, with education moving to a remote environment, to engage students, get them into programs, and keep them enrolled.

When asked if the tension around racial injustice triggered by the death of George Floyd was showing up in her work at Miles College, a historically Black college, Agee highlighted the new Miles College Institute for Economic and Social Justice, which will allow the institution to work with its partners, including its business sector partners, to define what moving forward looks like in terms of racial equity. She noted that many Miles College graduates become the first Black person to accept certain positions in some companies or institutions, whether in Birmingham or elsewhere, and they have the skills and talent to perform in that environment.

Responding to a question about creating authentic regional connections with employers to create pathways into careers and work-based learning experiences, Agee said that institutions of higher education need businesses to work together on building bridges that will help students succeed in the academic environment and then in the work environment. Benoit said that in addition to partnerships, she wants to see more businesses represented on college and university advisory boards to help develop curricula that are more responsive to businesses' needs and to serve as mentors to these students. Along the same lines, Anthony pointed to the importance of having education at the table in the early stages of regional business development and recruitment activities so that higher education institutions know what is coming and have time to develop programs to meet the needs of new and developing industries.

CLOSING REMARKS

To conclude the day's discussions, Carpenter; Maria Flynn, president and chief executive officer of Jobs for the Future; and Sarah McMillan, director of workforce and talent development for the city of Birmingham, summarized the key themes, challenges, and policy opportunities around workforce training and program development for Birmingham. McMillan commented that prior to the pandemic, Birmingham's metro area unemployment rate was under 4 percent, while the city's unemployment rate was nearly double that. By September 2020, roughly 54 percent of unemployment claims were for Black people in the city of Birmingham, even though Black workers only account for a quarter of the city's workforce. She called this a "concerning number that further illustrates the racial divide in the city, in our country, and a need for equitable and inclusive responsive work force and economic development initiatives, particularly in the city." Women were also disproportionately affected by the pandemic. Given that so much work has become remote and virtual, the digital divide is set to further increase the disparities affecting women and people of color.

Prior to the pandemic, the city commissioned the *Building (it) Together* report that highlighted the fragility of certain industries and the jobs that automation was putting at risk. COVID-19, said McMillan, has accelerated this prediction, particularly for jobs held by a large portion of the female and Black populations in the city (Bradley et al., 2018). The *Building (it) Together* report highlighted the need to create opportunities for more of the city's workforce to advance into higher-paying, family-sustaining jobs and the need for focused and intentional efforts to close equity gaps and promote access to jobs in growing industries.

McMillan then noted that today's sessions conveyed one crucial lesson. "When the pandemic first hit, city leaders believed that it was crucial to respond strategically to avoid losing time, because with each day that went by, the consequences of crisis sharpened. "Folks in Birmingham, the people that are directly impacted, cannot wait for us to create a novel, new, beautiful, fancy, shiny program. They need access to resources now," she stated.

She applauded the collaborative community efforts that enabled Birmingham Strong to place over 300 residents—mostly Black and female individuals—in various jobs, and Birmingham Promise's commitment to youth development and its investment in the marriage of workforce and educational experiences for young people that will ultimately shape their future and the future of Birmingham. She called out the collaborative efforts of Pack Health, Innovate Birmingham, and the area's higher education institutions to sustain a sense of long-term economic security and career advancement for the city's citizens, and credited the city's partnership with Birmingham Promise and Ed Farm with enabling the city to invest in its future citizens to create a prosperous Birmingham.

McMillan concluded, "Today, we learned that we are set up for success, and to achieve the city's longterm economic goals of creating an inclusive economy while offsetting the negative impacts of this pandemic, through our efforts in fostering workforce development, expanding economic opportunities for marginalized communities, and just creating a more robust community here in Birmingham. . . . we have the right people on the bus."

Flynn remarked that one thing she came away with after seeing the data that Andreason presented to open the workshop is that it is not just the unemployment rate that is important, but also the interplay between the unemployment rate and the poverty rate. That interplay means that workforce development efforts should not just be about getting people into jobs, but getting them into well-paying, family-sustaining jobs in an inclusive and equitable manner.

Reflecting on the first panel's message that resources are important, Flynn agreed that an increase in federal investment in workforce development is critical, but also suggested that innovative financing models are important given the likelihood that federal investment will not be enough. "Even in the best-case scenario where we see a dramatic increase in federal investment, there are still going to be gaps in resources," said Flynn. Vehicles such as career-impact bonds and income-share agreements may be useful, and Jobs for the Future is launching a financing-the-future initiative to look specifically at innovative financing models for workforce development.

From the second and third panels, Flynn took away the importance of having these various programs coordinate their activities and how they can help someone decide which of these programs would be the best one for their needs and skills. In that respect, she noted the importance of building an agile career navigation system. She also said that the digital divide that existed before the pandemic was exacerbated by workplace and school closures that required shifting to a more remote and digital world. This situation increased awareness about the lack of infrastructure and hardware for groups not prepared to work remotely. For the final panel her key takeaways were the importance of real-time labor market information and having the agility to adapt to changes in the labor market. In all, she said, she found the day inspiring.

Providing the final comments of the day, Carpenter agreed with Flynn that the interplay of unemployment and poverty was important, and so, too, is the interplay of labor force participation on race and gender dynamics. For him, the pandemic has underscored the fact that too many people of color and women are dependent on jobs that would not be considered quality or opportunity jobs. Addressing that disparity, he said, is going to continue to be the most important challenge facing communities like Birmingham.

His take from the employer perspective of the first panel was the importance of being nimble and adapting swiftly to a changing environment. Similarly, the second panel featured organizations that were also able to pivot quickly in the way they delivered their programs and scaled their efforts to reach more participants by going online and moving their case management model online. He thought that the idea of creating stackable credentials was interesting and wants to see more about how those are aligned with demand in the area. He was also impressed with the extraordinary efforts that the local colleges and community colleges made to move their programs online and continue to reach their students.

The final undercurrent that ran through the day, said Carpenter, was the use of federal funding. "We need to figure out how we are connecting to Workforce Innovation and Opportunity Act funding and maximizing CARES Act funding," he said, while also noting that Alabama was one of eight states selected for Reimagining Workforce grants to help people displaced by the pandemic identify work opportunities. He noted that while there is a great deal of action in the community, translating that action into an organized and systematic way of on-boarding folks who have been hurt by the pandemic disproportionately—women and people of color—should be at the forefront of these strategies going forward.

3

Boston, Massachusetts

In many ways, Boston, Massachusetts, represents a great science, technology, engineering, mathematics, and medicine (STEMM) success story embodying the revival of the American city over the past few decades, said Rachel Lipson, director of the Project on Workforce at the Harvard Kennedy School, in her introduction to the second virtual workshop of the six-workshop series. Prior to the pandemic, the city's economy was booming according to many measures, including historically low levels of unemployment, a thriving and diversified economy, and serving as headquarters to some of the nation's most innovative employers. Boston is the third-largest destination for venture capital funding in the country, with a globally renowned healthcare and life sciences sector and a technology sector that has grown rapidly in recent years. Boston's plethora of higher education institutions has created a steady stream of young people and STEMM talent, a significant share of which stay in the region after completing their degrees.

At the same time, the city's growth has not benefited all its residents, with wages for Bostonians without a bachelor's degree remaining stagnant for decades. In fact, said Lipson, Boston has one of the largest racial wealth gaps in the country, and inequity has been skyrocketing with a bifurcation between high-wage and low-wage job pathways. In addition, housing prices have made the city unaffordable for many, and gentrification remains a pressing issue. Moreover, the pandemic has exacerbated inequity across the Boston metro area, as many of the residents who were most affected by a loss of income were those who were already systemically disadvantaged.

Boston is a major tourist destination, with a normally thriving downtown corridor, but the pandemic has hit its hospitality industry hard. In addition, other sectors that have historically been countercyclical in times of recession, such as higher education and healthcare, have been affected given the unique dynamics of this pandemic. In addition, the lack of affordable childcare has exposed the challenges that working parents and especially mothers in the Boston metro area are facing during these challenging times. The varying level of reliable internet access in the region has also laid bare the stark digital divide that exists across the region.

Lee Wellington, founding executive director of the Urban Manufacturing Alliance, a national coalition of organizations and individuals that are building manufacturing economies fit for the 21st century, noted that addressing these inequities requires conversations that can be difficult at times, but the racial uprisings the nation is experiencing provide a pivotal moment to have those conversations and address inequities that are deeply embedded in STEMM workforce systems and their associated career pathways. "These are inequities that we must urgently correct in order to build resilient local and regional ecosystems and put ourselves on a concurrent path of scientific innovation and justice," said Wellington. She expressed hope that the day's discussions would serve to move the STEMM ecosystem along that path. In that respect, added Lipson, the Boston metro area can continue to attract young people and talent of all races, nationalities, and backgrounds, but ensuring that a STEMM-led recovery is shared broadly by all of Boston's residents is the defining challenge faced by the region.

This workshop began with sessions that presented data on economics, the labor force, educational attainment, and demographics for the Boston metropolitan area and the importance of creating an economic recovery that is equitable for Bostonians. The workshop then discussed how workforce development organizations, educational institutions, and training providers shifted their operations and program delivery mechanisms during the COVID-19 pandemic; the ways in which the pandemic has exacerbated existing inequities in the region and the ongoing work to embed anti-racism in training activities; and the linkages between STEMM workforce development and entrepreneurship.

UNDERSTANDING THE ECONOMIC AND WORKFORCE IMPACTS OF COVID-19 ON BOSTON

To ground the day's discussions, Lipson presented a wealth of data from the Atlanta Federal Reserve Center for Workforce and Economic Opportunity; the Bureau of Labor Statistics; and Opportunity Insights, a non-partisan, not-for-profit organization located at Harvard University that seeks to translate insights from rigorous, scientific research to develop scalable solutions that will empower people throughout the United States to rise out of poverty and achieve better life outcomes. While median wages were only slightly lower in the city of Boston than in the greater Boston metro area in 2019—$45,046 and $49,592, respectively—the poverty rate in the city was nearly double that of the metro area, at 17.1 percent versus 8.6 percent. Lipson noted that between 1980 and 2017, Boston lost middle-income households, while the numbers of high- and low-income households both increased, illustrating the high level of income inequality seen in Boston and many other U.S. cities and the fact that Boston's economic boom has not translated equally across the city's population. Boston, she explained, is a majority minority city, with larger shares of Black and Latinx residents than the metropolitan area (Figure 3-1).

Regarding educational backgrounds, the greater Boston area has one of the highest rates of college graduation in the nation (Figure 3-2), with almost half of the residents of the metro area having a college degree compared to 33 percent nationally. The region also has almost double the share of workers with graduate or professional degrees compared to the national

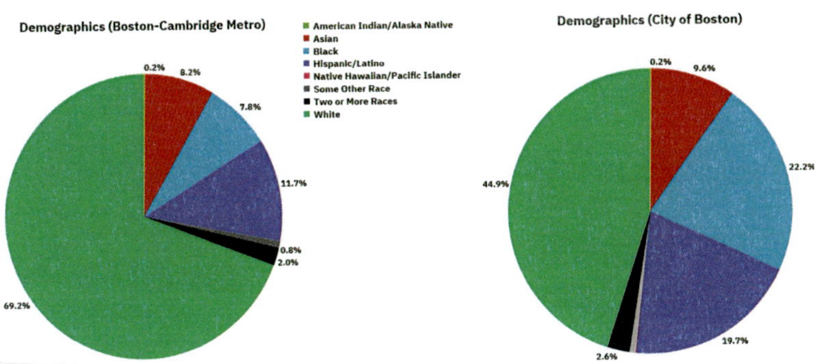

FIGURE 3-1 Demographics of the greater Boston metropolitan area (left) and the city of Boston (right).
SOURCE: Lipson slide 3.

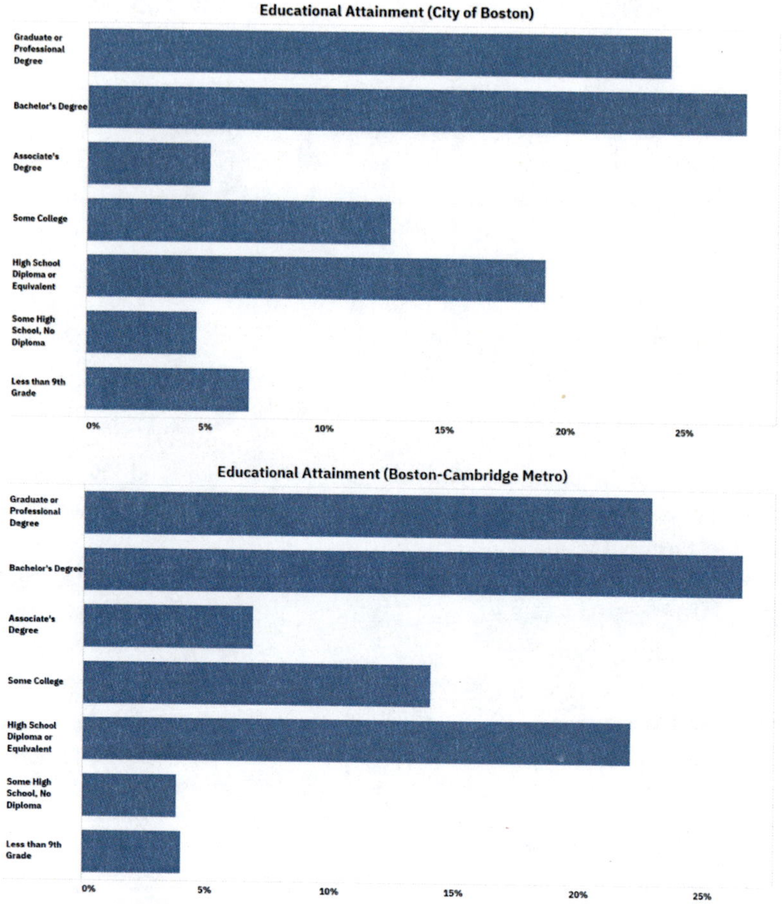

FIGURE 3-2 Educational attainment in the city of Boston (top) and the greater Boston metropolitan area (bottom).
SOURCE: Lipson slides 4 and 5.

average. Educational attainment in the city proper is similar to that of the metro area except there are higher shares of residents in the city who do not have a high school diploma or who have less than a ninth-grade education.

Boston and its metropolitan area have seen a massive spike in unemployment that started in April 2020. While the unemployment rate has been declining since June 2020, it is nowhere near pre-COVID-19 levels. Unemployment since the start of the pandemic has disproportionately affected women, adults under age 35, and Black or African American workers in Massachusetts. Women, for example, made up 55 percent of those who claimed unemployment as of September 2020 while making up just 45 percent of the workforce, and Black or African American workers accounted for 12.6 percent of the claimants while making up just 8.7 percent of the workforce (Figure 3-3). Lipson noted that as of August 21, 2020, low-wage jobs in greater Boston—those paying less than $27,000 a year—had declined by 25.3 percent, compared to 16.2 percent for middle-wage jobs and 5 percent for high-wage positions (those paying more than $60,000 per year). Similarly, for the week ending October 9, 2020, postings for jobs requiring "minimal" education had fallen by 52.9 percent from January's level compared to a 30.2 percent decline for all job postings and a 10 percent decline for job postings requiring "extensive" education.

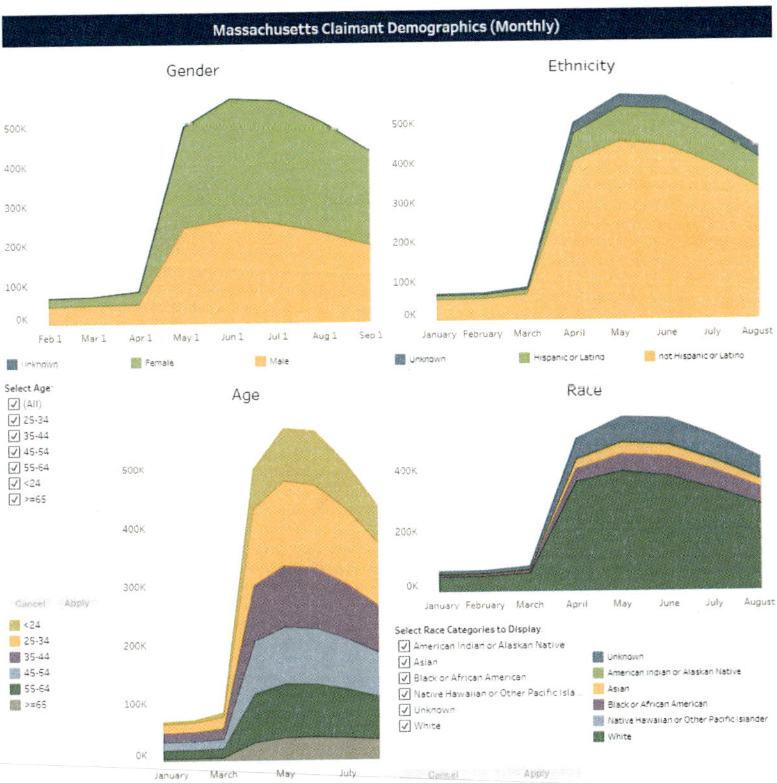

FIGURE 3-3 Unemployment claims for Massachusetts by age and race.
SOURCE: Lipson slide 8.

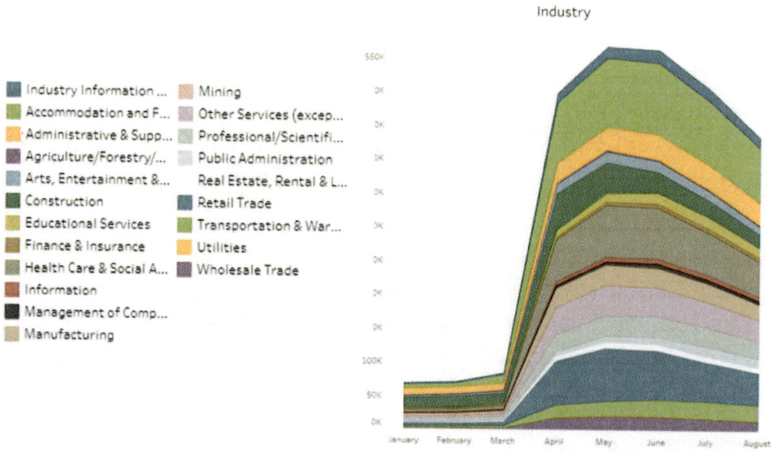

FIGURE 3-4 Unemployment claims for Massachusetts by industry.
SOURCE: Lipson slide 12.

Statewide, all job postings had fallen by 11 percent, compared to a nationwide decline of 16.3 percent over the same period.

By industry, the largest share of unemployment claims was from individuals in the accommodation and food services sector of the statewide economy, followed by healthcare, retail trade, and administrative support and waste management, sectors that disproportionately employ women, minorities, and lower-income workers (Figure 3-4).[1] In addition, the Atlanta Federal Reserve Bank estimates that Boston's opportunity employment situation—defined as jobs that do not require a bachelor's degree but pay above regional median wages—is less strong than for other parts of the country.

Lipson discussed how the mix of industries in a region has a direct bearing on the number of jobs that could be lost in that region as automation continues to take hold across different employers. He noted that the Federal Reserve Bank of Atlanta's analysis of regional risk of job loss from automation suggests that Boston, like many urban centers, ranks at the lower end of the spectrum, suggesting that big, diversified economies are best prepared to deal with automation. This is consistent with estimates by the Congressional Budget Office that automation will affect 41.1 percent of

[1] Unemployment claims by industry were only available by state, and not by city or metropolitan area.

greater Boston's industry employment demand, compared to 42.7 percent for the entire state (Arnold et al., 2006), while the Atlanta Federal Reserve Bank estimates that 30.9 percent of Boston's workers will be able to work at home, compared to 39.9 percent of Massachusetts's workers (Andreason et al., 2020).

Data from the Census Bureau's American Community Survey revealed that 81.2 percent of households in the greater Boston metropolitan area have a high-speed internet subscription, 8.1 percent access the internet through cellular data plans alone, and 9.5 percent have no internet subscription. Nearly one-third of households earning less than $20,000 annually have no internet subscription. For computing devices, 83.2 percent of all households in the metro area have a desktop or laptop computer, 6.2 percent have a smartphone and no other device, and 5.7 percent have no computing device. Those figures are slightly lower in the city proper, with almost 10 percent of households accessing the internet solely through their smartphones, over 12 percent having no internet subscription at all, and 7.3 percent of all households having no computing device.

A RECOVERY WITH EQUITY

Trinh Nguyen, director of workforce development of the city of Boston, said that if racial equity and the integration of STEMM and workforce development was bad pre-COVID-19, it is worse now. She noted that her office is the largest regional funder of workforce development efforts in the region, funding between $14 and $20 million annually on programs run by its partners, including community colleges and nonprofit organization in the area. While she acknowledged that this is hardly enough money to have a big impact on workforce development, it enables her office to leverage other resources to provide opportunities such as tuition-free community college.

In terms of workforce development and STEMM, Nguyen's office focuses on 18- to 24-year-old adults, including English language learners, who do not have a bachelor's degree. Most of her office's clients are people of color, single women heads of household, and families and individuals that have multiple economic barriers. She explained that for workforce development for these underserved groups to be considered successful, it cannot include only education and training, but should also include a job at the end of the process. Therefore, employer engagement is a critical piece of their efforts.

One example of those efforts is the pre-apprenticeship program developed by the city of Boston's Office of Workforce Development in collaboration with 13 building trades across the Boston region. At the time of the workshop, some 420 people of color, women, seasonal workers, and others were enrolled in the program as an entry point into a career in the building trades. Nguyen noted that the program has generated a great deal of interest among people of color who are taking the opportunity to train to become electricians and pipe fitters and enter other technical trades.

Her office also partners with the Boston Private Industry Council, which provides skill training and internships for young people from diverse groups to enable them to gain experience in and entry into the region's tech industries. Another important source of STEMM employment is the city itself, which has various positions requiring engineering-type training, such as building and infrastructure maintenance. The City Academy Program provides training for and a pipeline directly into those positions. Another program provides emergency medical technician training, which includes the state's first registered, paid apprenticeship program that guarantees that graduates will have jobs when they complete the program.

In closing, Nguyen said that when it comes to STEMM and equity, one major lesson she has learned is that if there is an opportunity for training and jobs, people of color will obtain those competitive credentials, get into those fields, and do well in them. At the same time, however, she and her colleagues have found that some employers hiring people of color who have completed these programs still start them in entry-level jobs, rather than the jobs they are now qualified for. Even more concerning, many graduates are kept in those entry-level jobs and not given the opportunity to move up in their fields. She believes that some hard conversations are needed to address this problem.

COVID RESPONSE AND WORKFORCE TRANSITIONS IN GREATER BOSTON

The first panel session highlighted how workforce development organizations, educational institutions, and training providers have shifted their operations and program delivery mechanisms in the pandemic environment. Topics discussed included emergency and contracting STEMM workforce development opportunities in greater Boston, partnerships to support operations in a resource-constrained environment, and continuity in training and wrap-around services as programs move online. The panel featured four speakers: Kaitlyn Bean, senior program officer at the Boston Foundation; Lane Glenn, president of Northern Essex Community College;

Joanne Pokaski, senior director of workforce development at Beth Israel Deaconess Medical Center; and Jerry Rubin, president and chief executive officer of Jewish Vocational Service Boston. After the four presentations, Deeneaus Polk, Bosch Fellow at the Robert Bosch Foundation, moderated a question-and-answer session with the panelists.

Working with the Most Vulnerable Communities

Bean noted that three simultaneous crises—the COVID-19 pandemic, the resulting economic crash, and the racial reckoning triggered by George Floyd's death—are having a profound impact on workers and families, particularly among low-income individuals and communities of color. Prior to the pandemic, her organization's SkillWorks initiative[2] and its partners were exploring strategies to leverage greater Boston's tight labor market to drastically change how workforce development is done. With the pandemic raging, this activity became what she called a moral imperative. Since March 2020, SkillWorks has been in close contact with all of Boston Foundation's grantees and other partners to conduct listening sessions, share information, and determine how the foundation, as a funder, convener, and influencer, can best support the immediate needs of the community and greater Boston's overall economic recovery in the months and years ahead. As an early response, SkillWorks converted all of its funding to general operating support and suspended all reporting requirements in order to reduce as much administrative burden on its grantees as possible.

Noting that her organization's grantees were already working with the most vulnerable communities before the pandemic hit, Bean said it has been inspiring to see how they have sprung into action and transitioned to remote service delivery. On top of that transition, these organizations have had to help massive numbers of clients with an array of other needs and challenges brought on by the pandemic, including navigating the unemployment system, gaining access to computers or mobile devices and reliable internet, avoiding eviction, and accessing cash supports in emergency situations. She added that these services were not just for current or new clients, but alumni of the program who are also now struggling.

To help support frontline staff, SkillWorks continued its career coaches' community of practice by offering virtual learning sessions on topics such as

---

[2] Additional information is available at https://www.tbf.org/what-we-do/special-initiatives/skillworks.

digital literacy and transitioning to remote learning, doing career coaching remotely, supporting clients who are experiencing trauma, and supporting staff who may be experiencing trauma and toxic stress themselves. Bean's hope is that federal and state money will eventually help support this work.

Community Colleges Respond

Glenn described how Massachusetts has 15 community colleges that serve some 150,000 students annually, accounting for 49 percent of the state's undergraduate enrollment. In 2019, the community college system's students earned more than 10,000 associate degrees and more than 3,300 certificates. At Northern Essex Community College, one-third of the graduates received their degrees and certificates in a STEMM field.

When the pandemic first hit, Glenn wrote an article asking how community college students differ from other students (Glenn, 2020). In general, community college students tend to be lower-income students, and it is not unusual for more than half of the students on a community college campus to be the first in their families to attend college, he explained. They also tend to be older, people of color, working—sometimes multiple jobs—while attending college, and often raising families. Every year, approximately one-third of Massachusetts's community college students report being without housing or experiencing food insecurity, and they tend to be less connected in terms of having computers and broadband internet access at home.

His institution, and likely all of the other community colleges in the state, announced early during the pandemic that it planned to move its courses online for the fall 2020 semester. Making that decision early allowed the community college to spend the summer training faculty and redesigning some courses to be ready to provide online education for its students. Glenn noted that funding from the Coronavirus Aid, Relief, and Economic Security (CARES) Act provided to colleges and universities helped his institution pay faculty to convert their courses for online delivery and put the equipment in place, including providing wireless internet access in all of its parking lots. His institution also implemented a laptop requirement for its students and provided resources to make sure every student would have a laptop computer if they did not already have one. To support students who might have issues with online learning, he and his colleagues also spent the summer training student ambassadors to help with that type of issue.

Among the struggling occupations that community colleges tend to train people for, hospitality and food service took big hits, an unfortunate development as the economy crashed the very month his college was opening the multi-million-dollar Hospitality and Culinary Arts Institute it had been planning for years in response to industry demand. Glenn said that his institution, as well as many of the state's other community colleges, also offers an associate degree in early childhood education, another area that has been hit hard by the pandemic. Jobs in other fields where community colleges commonly provide training have also been heavily affected, including healthcare fields such as radiology technician and dental hygienist and some areas of manufacturing. There are bright spots, though, for jobs in cybersecurity and information technologies, as well as for laboratory technicians, licensed practical nurses, and medical assistants, for which demand has stayed relatively strong. On a final note, Glenn described how the community colleges are putting out an "innovation playbook" on promising practices for fostering innovation in post-pandemic health education.

Healthcare Training Programs Adapt

Pokaski said that the Beth Israel Deaconess Medical Center has been heavily affected by the pandemic and has had to put trainees and practicums on hold as it dealt with the enormous medical emergency wrought by COVID-19. The process of bringing in new trainees, even for high-demand roles, had to stop given that a pandemic was not a good environment for bringing in new people. Her institution also paused its program that sponsored people with an associate degree in nursing to get their bachelor's degree while working for the medical center. She explained that while this is a great program, managers were too overwhelmed dealing with the medical crisis to interview candidates for the program. She also noted that many employees who had jobs such as welcoming people or helping them navigate the medical center were redeployed to help with COVID needs.

At the time of the workshop, Pokaski said her institution had switched out of incident command mode and had established a plan to handle a possible second surge of the virus. While classes for employees will remain virtual through the year, the medical center was working to fill over 600 vacancies, particularly those for patient care technicians and medical assistants. Training programs had restarted, and the medical center is marketing them digitally and conducting Zoom job interviews. It has also resumed practicums.

Pokaski ended her presentation with some policy recommendations. One was to consider healthcare roles with high demand and low barriers to entry for those needing quick job transitions. The second, directed at healthcare educators, was the need to convey a better understanding of the role of the healthcare worker in an emergency so that prospective students will be more informed before entering a healthcare field. A third recommendation was for the state to find creative ways to bring back testing and certification capacity so that people who are finishing programs are not held back waiting for a test and unable to move into jobs immediately. Finally, she would like to see more support for jobs seekers with limited technology and technical skills. She wondered whether Boston could use this moment to lean into training more people for information technology roles.

New Roles for an Experienced Organization

In his remarks, Rubin noted that Jewish Vocational Services was founded in 1938 and has therefore gone through one depression and one great recession, and it has operated in great job markets and horrible ones. In a typical year, it partners with dozens of employers and provides both pre-employment services and incumbent-worker career ladder services to 15,000 to 20,000 people per year. In 2020, however, the overnight collapse of the job market required the organization to shift to focus on its clients' immediate needs, including helping them with unemployment, re-employment, and emergency cash assistance for the first time ever. He explained that while the state did a good job helping the suddenly unemployed, the system was overwhelmed. In addition, many of their clients, particularly non-native English speakers, had a hard time navigating the unemployment system.

Like the other speakers' organizations, Rubin's shifted immediately to providing its services remotely, offering online classes using Google classroom, Zoom, and WhatsApp, which many immigrants were comfortable using; setting up a laptop and wireless hotspot lending library; and implementing volunteer "digital navigators" to help those clients who had limited digital literacy. It also developed and launched a remote talent match portal for online employment services so that employers could post jobs in a curated manner and job seekers could upload resumes and work with career counselors.

At the time of the workshop, Rubin said that Boston was experiencing a strong and robust healthcare and life science employment market, and though his organization has not slowed its training in the healthcare area, there is the current problem of limited capacity for testing and certification mentioned also by Pokaski. Manufacturing, retail financial services, and administrative positions were recovering, and remote distribution service jobs were in high demand. The hospitality industry, however, remained in crisis mode.

Rubin's policy and strategy recommendations included expanding resources for proven employment and skills services and expanding investments in remote learning and remote employment services. He also suggested improving the availability of real-time job growth and hiring data and continuing a focus on job quality within the context of a softer job market. Finally, he called for expanding proven programs for delivering workforce development services and quality jobs in Black and brown communities.

Discussion

Speaking about the nature and availability of jobs in greater Boston since the onset of the pandemic, Pokaski said that while the hospitality sector has struggled, there are hospitality jobs available in the healthcare sector. She believes there is an opportunity to help people transition from hospitality to healthcare. Rubin echoed Pokaski's comment, noting that the chief executive officer of one of metro Boston's many hospitals told him that healthcare organizations in the region serve more dinners and have more beds than Boston's entire hospitality industry. Rubin also said that the local job market in the interconnected fields of healthcare, life sciences, and medical technology is stronger than ever, with opportunities to move from one field to another and for people to reposition themselves as the demand for jobs ebbs and flows among the three fields.

Glenn called that situation an adjacency, an opportunity to take skills from one sector to another or, in the case of his institution, renting out space in its new Hospitality and Culinary Arts Institute building to home-based business and entrepreneurs who might not otherwise have access to a commercial production space. These types of connections create the opportunity to create new businesses or make new connections between different sectors of the economy. "We are taking a closer look at on ramps and how people discover new fields in the first place," said Glenn. He added that as fields expand, the goal should be to expand the pipeline into those fields.

Bean, whose organization works with the hospitality, healthcare, and information technology sectors, said that the organization's information technology training partners were able to get back up and running quickly, have maintained their relationships with their employer partners, and are placing people in jobs. However, its partners in the hospitality training program reported that over 90 percent of the local hospitality sector's employees were laid off within the first month of the pandemic. In response, one hospitality training program pivoted quickly to start training laid-off housekeepers for environmental services positions at hospitals. One barrier she noted was that people with a criminal background find it easier to get a job in the hospitality industry and move up in that industry than at a hospital. Another barrier, said Bean, is implicit bias and racism.

When asked about opportunities to reset how the workforce development system engages or trains people, Rubin said that his organization's one-stop career center now operates remotely, and he sees no reason why it will not continue to operate remotely once the pandemic is over. In fact, he sees no good reason why much of the service delivery that the workforce development system does now should not be done remotely, given that it makes accessing these programs easier and take less time. "We are delivering exactly the same content for individuals that need to maintain their unemployment insurance and meet certain criteria to maintain that insurance, and we are having a higher level of participation," Rubin explained. "Why should we not continue to make things easier for consumers?" His organization has moved all of its recruiting and assessment activities online, and the result is an increase in enrollments. He acknowledged, though, that remote learning may not be the preferable option for everyone, particularly for individuals with lower-level English language skills.

Glenn agreed that there is no good reason not to keep delivering much of his institution's classes and services online while also acknowledging that some course offerings, such as cardiopulmonary resuscitation training for paramedics, will still have to be delivered in person. That said, his institution is looking at how to make its offerings available to a broader population of students and how to use technology to shorten the time needed to earn a credential. In the same vein, Bean pointed out that the pandemic has created the opportunity to recognize that whatever was normal in the past was not working for a large portion of the population, particularly people of color. She did note the challenge of preparing people for jobs and careers while also anticipating that a chosen career path might end at some point in the future.

When asked if there were any complications around higher education partnerships for building STEMM-based skills, Glenn said that technology has lowered barriers to successful partnerships among higher education, training programs, and employers. As an example, he cited a program that Bean's organization is funding among the state's community colleges to implement a product called Handshake that will help employers connect directly with students either upon graduation or as they search for internships. Pokaski added that finding the right partner in a higher education partnership and working out the kinks can be a little challenging, but it is smooth sailing after that, and Rubin pointed to the importance of all parties recognizing the value that they bring to the table as a key to developing robust partnerships. Bean noted that money always helps develop stronger partnerships, particularly when it comes to helping students pay for books, transportation, childcare, rent, and other expenses they incur while training or for providing laptops and wireless connections.

COVID-19 AND BOSTON'S ECONOMIC DIVIDE: THE INTERSECTION OF THE PANDEMIC, RACIAL AND SOCIOECONOMIC INEQUALITY, AND ACCESS TO GOOD JOBS IN STEMM

The second panel focused on ways that the COVID-19 pandemic has exacerbated existing inequities in Boston and how multi-stakeholder approaches to STEMM workforce development can directly address those inequities. The panel's four speakers were Tyra Anderson-Montina, Boston site director of Year Up, a program that connects young adults and companies to launch careers, power business, and build community; David Delmar Sentíes, founder and executive director of Resilient Coders, which trains people of color for high-growth careers as software engineers and connects them with jobs; Valerie Roberson, president of Roxbury Community College; and Neal Sullivan, executive director of the Boston Private Industry Council, which works at the intersection of business and community interests to connect Boston residents to promising career pathways while creating a diverse talent pipeline for local employers. The four speakers explored how different institutional actors are working to embed anti-racism in their training and pipeline development activities. They also commented on how their institutions adapted their programming in response to the death of George Floyd and the subsequent racial uprisings that occurred across the nation. Following the four presentations, Peter Blair, assistant professor of education at the Harvard Graduate School of Education, moderated a discussion with the panelists.

Connecting Inner City Community College Students to City Services

Robertson explained that Roxbury Community College is located in the center of Boston and serves some 3,000 students a year, of whom 90 percent are students of color and 80 percent are Pell grant–eligible. These students, whose average age is 30, are at the center of the pandemic's impacts, said Roberson. In general, her institution's students are the city's essential workers, and when the pandemic hit, many experienced an interruption in their studies because of the loss of income. Even before the pandemic, these students' studies were often complicated by issues concerning their health (pre-existing conditions that would make them more vulnerable to COVID-19), and regarding insurance coverage, access to healthcare, access to grocery stores, food insecurity, and housing insecurity.

To address as many of those obstacles as possible, Roberson and her colleagues have made a conscious effort to be more proactive about staying in closer contact with their students, making sure they were well and helping them access services throughout the city through a program called Project Access that makes connections between her institution and city services. Roberson noted that her school is a reflection of issues its students have, including being chronically underfunded. Fortunately, the business community has been a good partner and has helped develop and fund new programs, has provided opportunities for internships, and is providing employment opportunities for students. Going forward, Roberson is hopeful that the pandemic will bring about new ways to partner and to be more effective in creating schedules that help students be successful.

Investing in Public Schools

Sullivan noted that one of the things that makes Boston strong is that there are several interlocking boards and initiatives that work together regularly, which has allowed the city's workforce development efforts to be as nimble as possible in these unprecedented times. The Boston Private Industry Council serves as the city's workforce development board, one of 16 in the state. In that role, it oversees the city's two career centers, one managed by Rubin's organization, the other by Morgan Memorial Goodwill Industries.

Starting in 1982, he explained, his organization decided that the Boston public school system would be its baseline investment. There are now career specialists in Boston's high schools, each of whom works with some 150 students to identify, prepare students for, and match students with jobs and internships, and an employer engagement team that works with greater Boston's major employers, including its largest, the region's hospitals. In addition, there are re-engagement specialists who "chase down" students who have dropped out of school. A big issue for his organization when the pandemic struck was what to do about the thousands of jobs and internships that it had arranged for students. It took a heroic effort to move everything online, an effort that was helped as employers also pivoted to online working and virtual learning programs and internships.

Providing Hands-on Training Opportunities

Anderson-Montina explained that Year Up, which serves students aged 18 to 24, has been operating in greater Boston for some 20 years. Its mission is to close the opportunity divide by providing a hands-on skills curriculum, training, and brokering opportunities with employers for internship placements that serve as on-the-job training opportunities. This approach, she said, empowers students to find themselves on a journey toward a stable career and higher education attainment. The challenge that the pandemic triggered for Year Up is that it is a "high-touch" organization in how it works with its students and provides them with wrap-around services; therefore, figuring out how to do that virtually was not an easy feat, particularly given that the digital divide is real for most of its students. In addition, as Roberson noted, the pandemic has increased food and housing insecurity among the students, in addition to taking a toll on their mental health, particularly considering the racial trauma that so many of the communities it serves have experienced over the past year.

Coding Bootcamps to Address Equity

Delmar Sentíes explained that Resilient Coders is a highly competitive, free coding bootcamp for young adults of color who come from low-income backgrounds. Some 85 percent of its 2019 graduates found full-time jobs—with an average salary of $98,300—within a month of completing the bootcamp, and there are more commitments to hire than there are students in the program. His organization's mission is to recalibrate power in America, and Boston specifically, by creating pathways to college that do not necessarily depend on going to college, which is just not an option for many of its students. As he put it, "We must convene a league of equity-led organizations and employers, and thrust into the mainstream our 'radical' ideas about social justice in technology."

Achieving that goal, he said, requires five actions:

- Develop a body of principles and agitate for their adoption. Toward that end, Delmar Sentíes believes that Black and Latinx technologists must codify what it means to employ and advance equitably, and allies in these efforts must hold their leadership accountable to their publicly stated values.
- Move past diversity and toward equity and justice as a means of recalibrating power. Hiring a person of color into a dead-end job is an example of diversity without power.
- Rethink education and come up with a way to make top-tier education not just affordable, but free as a civil right and as a means of countering the stratification of wealth in America. Requiring people to pay for their education helps to maintain the stratification of wealth in this country.
- Create a league of equity-led organizations and employers and develop a body of principles crafted by Black and Latinx technologists. One piece of such a body of principles should be that employers do not require a bachelor's degree for jobs that do not require a bachelor's degree to perform, such as coding.
- Overhaul the way workforce development is funded. Currently, most funding for nonprofit workforce efforts comes from philanthropy and leading capitalists who want to ameliorate the harmful effects of capitalism on communities of color. In Delmar Sentíes's opinion, there is a need to find a more democratic way to fund workforce development, perhaps by a corporate equity tax on those companies that do not abide by the body of principles.

Discussion

Regarding wrap-around services and their importance for education, Roberson noted that Roxbury Community College has a partnership with Year Up in which the college provides the education and Year Up handles all the wrap-around services, including a monetary stipend and training in soft skills. Providing wrap-around services is particularly important for community college students who have a much lower completion rate than those who attend four-year institutions. "This is not a situation where our students are less intelligent," said Roberson. Rather, it is a variety of barriers. For example, community college students progress at half the speed of four-year college students, so providing assistance helps them persist and complete their education.

Anderson-Montina seconded how important it is to provide these wrap-around services that can support students through completion of their education. During the pandemic, it has become clear, she said, that more mental health services are needed, and perhaps telehealth options can help there. Year Up is trying to broker a partnership with an app-based organization for people of color.

Delmar Sentíes pointed out that he does not have to convince Black and Latinx young adults to get into coding, for they already know that coding ability translates into well-paying jobs. What he does have to do is spend time educating employers about how the result of the systemic racism pervading their recruitment practices—specifically, a non-inclusive company culture—can cause some of their Black and brown employees to leave their companies after only a short time.

When asked about the linkages that need to develop between community colleges and four-year institutions to deal with racial justice in Boston, Roberson said that it is important that any time a college creates a degree program, it should consider both what the student will do afterward and what the student needs to get there, which taken together will dictate the partnerships that need to be arranged. For example, Roxbury Community College has partnerships with area high schools to help students who want to go into STEMM fields determine the courses they need to be successful in college and then have the high schools offer dual high school/college credit for certain courses. On the other end, her institution works with employers so that when these students complete their studies, they will have jobs to enter.

In terms of takeaways from the session, Sullivan's lesson learned was that people do not have to be physically together to learn together. "[The online] environment allows us to connect with people far beyond our personal associations or even our own community and see different worlds quickly," said Sullivan. Delmar Sentíes said the pandemic has provided an opportunity to organize, unite, and push for change in the way that the industry recruits, hires employees, and advances. Anderson-Montina said that it is imperative not to overlook talent any longer and to make opportunities inclusive for everyone. Roberson said the pandemic has highlighted the interdependence that may have gone unrecognized before the pandemic. A recognition of interdependence argues for no longer ignoring groups of people who are the essential workers keeping society running. "The pandemic has highlighted that in order for anybody to succeed, we all have to succeed, and there is hope in recognizing that fact and working together to solve it," she said.

CATALYZING STEMM ENTREPRENEURSHIP FOR ECONOMIC MOBILITY

The final panel of the day explored the linkages between entrepreneurship and STEMM workforce development and articulated how entrepreneurship can provide new routes to economic mobility for Bostonians once the pandemic-triggered crisis ends. As noted by Lee Wellington of the Urban Manufacturing Alliance, entrepreneurship can itself be a workforce development system, given that that many careers can emerge through entrepreneurship, often creating wealth, without the same credentials and requirements that traditional workforce pipelines might require. Specific topics in this session included how entrepreneurship can create job transition pathways, provide insights into emerging STEMM careers, catalyze ownership and employment opportunities, and provide a foundation for workforce education. This discussion also highlighted how entrepreneurial support organizations—including incubators, accelerators, and makerspaces—have worked closely with community-based partners to build ecosystems with strong connections to STEMM workforce development programming.

The three speakers for this panel were Jen Faigel, executive director and co-founder of CommonWealth Kitchen, Boston's food business incubator; Sonia Moin, senior director of urban business initiatives at the Initiative for a Competitive Inner City, a Boston-based national nonprofit organization that drives inclusive economic prosperity in under-resourced communities through innovative research and programs to create jobs, income, and wealth for local residents; and Michelle Weise, senior advisor and entrepreneur-in-residence at Imaginable Futures, a global philanthropic investment firm focusing on driving innovations that holistically support learners, from academics to life skills, and enable lifelong learning and lasting wellbeing for them, their families, and their communities. Moderating the discussion was Cecelia Wessinger, principal and founder of Mass Collaboration, an organization focusing on entrepreneurship as a means of economic development.

Matching Local Demand for Specific Skills

Weise said that each region of the country has what she calls a skill shape that describes the demand for specific types of skills in that region.[3] For example, in southern California, with its concentration of defense contractors, there is a demand for more traditional manufacturing skills such as welding and machine skills, whereas in northern California, with its concentration of technology companies, the manufacturing jobs there demand more STEMM skills (Figure 3-5). Similarly, said Weise, digital marketing roles in Denver, Colorado, are oriented toward analytics and cloud computing, while in Boise, Idaho, they are oriented toward design and web development and in Atlanta, Georgia, they are geared to marketing automation (Figure 3-6). These regional differences in job skill demands are important to understand for job seekers who want to work in a certain region and need to know what skill gaps to fill to make themselves competitive for jobs.

What is important to understand, said Weise, is that individuals also have skill shapes that represent the different assets and unique skills they bring to the table. Comparing a collection of individual skill shapes to the type of skills an industry requires can help identify skills where the supply is not meeting the demand and skills where the supply outstrips the demand. "As we think about advancing and mobility, especially as we face this tough economic recovery, we have to think about how we start to fill some of those gaps and the ways in which we can connect to educational programs," said Weise. This information about gaps can also point to ways in which regions can help people transfer their skills and pivot from one industry to another.

Weise explained that another way to use data to help people move from one industry to another is to map out where people go from retail and customer service jobs, for example, into many other fields, and then identify the skills they acquire that enable them to move to better-paying jobs. These newly acquired skills can then inform the types of micro-credentials and certificates that educational institutions might want to offer to students in their region. Weise said, "As we think about recovery and mobility for the future, especially for the bottom quartile of our population, we have to

[3] Typically, skill shape encompasses two dimensions: one axis represents the depth of related skills and expertise in a single field or multiple fields, and the second represents the ability to collaborate across disciplines with experts in other areas and to apply knowledge in areas of expertise other than one's own.

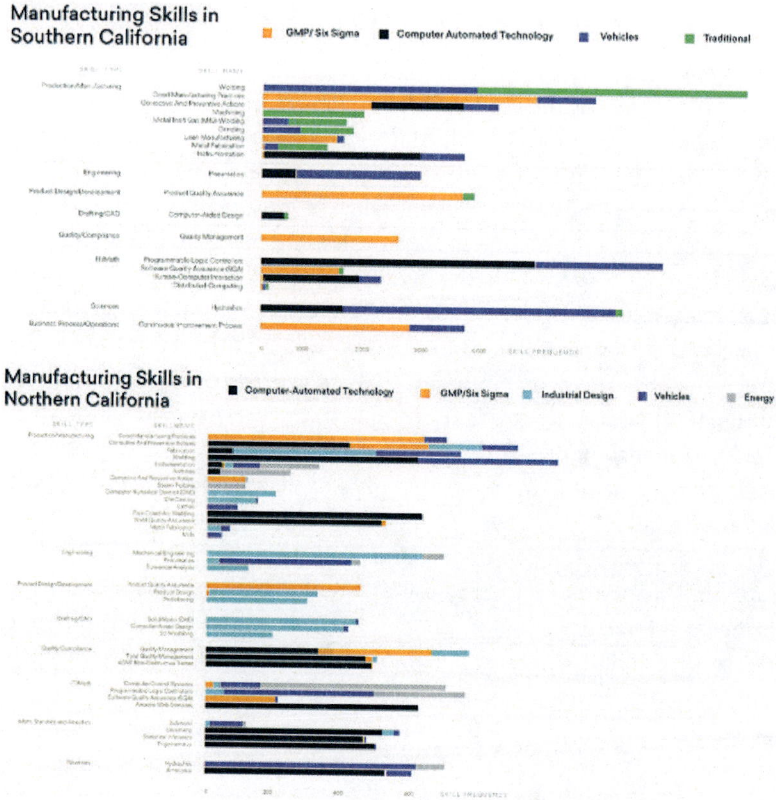

FIGURE 3-5 Skill shapes for southern (top) and northern (bottom) California illustrating the different skill sets needed for regional manufacturing industries.
SOURCE: Weise slides 3 and 4; reprinted with permission of Robert Sentz.

think about how we help people move from survival mode into something that can lead to better economic opportunity with just a handful of skills they can build quickly."

Becoming an Entrepreneur

Faigel addressed what it is like to be an entrepreneur who had been in a low-wage job with no potential for advancement. Her experience at CommonWealth Kitchen has shown that a person's confidence level soars even as they are just starting their business and start making connections and forming their own networks. They start to understand that they have

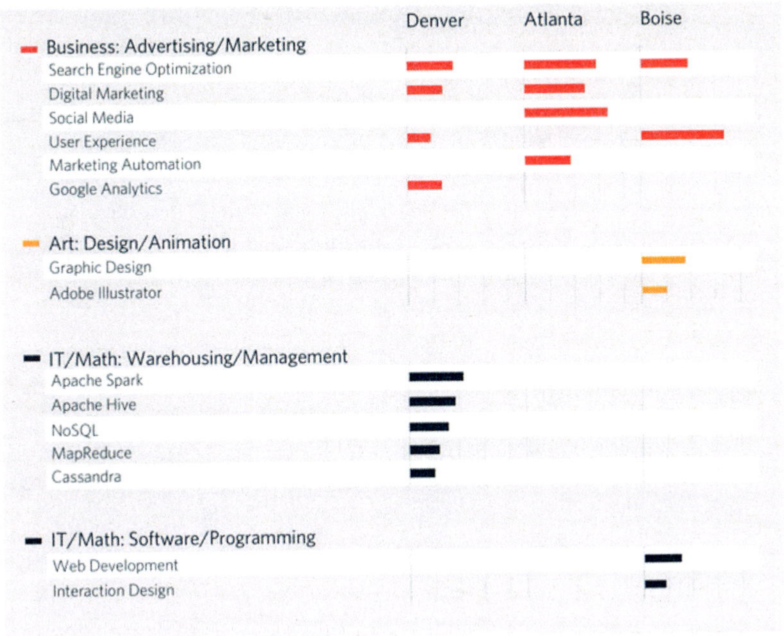

FIGURE 3-6 Skill shapes for digital marketing roles in Denver, Atlanta, and Boise. SOURCE: Weise slide 7; reprinted with permission of Robert Sentz.

useful skills that with some extra training they can use in new ways to build a better life. She noted that she spends a great deal of time with her entrepreneurs helping them understand what success is in the food industry. What Faigel would like to understand better is how this newfound confidence and resilience puts people on a different course in life, even if their first attempt at starting a business fails.

When asked what advice she would give organizations that support entrepreneurs, Moin replied that when wrap-around service providers think of creating a program or support service, they should consider locating the program so that it is easy for the targeted business owners to access. If the program will be virtual, it is important to know whether people in the targeted demographic will have access to a computer and reliable internet connection. "I know we would love to believe that if we build a program, they will come and we will help everybody, but that is not the case," said Moin. "If you are going to build it, you as the provider have to go where they live, where these entrepreneurs are going to be. You have to go to those underserved areas to implement your program."

Enhancing Sustainable Entrepreneurial Opportunities

When asked what they would do if they could wave a magic wand to enhance sustainability and create robust and thriving opportunities for people, Weise replied that what is missing from the entrepreneurial market are growth accelerators that will support people who get through the early stages and now want to scale up their businesses and get them to a sustainable size. Currently, she said, most of the funding emphasis is on early-stage entrepreneurs. Both Faigel and Moin supported this idea, with Moin noting that one of the programs she works on, the Goldman Sachs 10,000 Small Business program, was created specifically to support those businesses that want go from having four full-time employees to 10 full-time employees, for example.

Faigel implored cities and investors to stop using their money to entice big companies such as Amazon, which Wessinger noted do not create that many new net jobs, to participate. Instead, she would like to see them invest both financial and non-financial resources in the small businesses that want to expand in their communities and that truly create new net jobs. As a concrete example, she cited how much of the food feeding Boston children who are learning remotely during the pandemic is prepared in New Jersey by the food services division of a California-based company. Instead of taking the $20 to $30 million dollars that the federal government is sending to Boston, which then spends it in New Jersey where the money eventually flows to a multinational corporation based in California, that money could be invested in local businesses that would employ local workers and support local farms and fisheries.

Faigel noted that CommonWealth Kitchen is part of a coalition, the Massachusetts Equitable Paycheck Protection Program Coalition that went to local banks and told them they need to work with Black-, brown-, and immigrant-owned businesses and spoke to the business owners to convince them that they needed to apply for Paycheck Protection Program funds. The coalition then advocated with the Small Business Administration to forgive these loans for small businesses, which she said appears to be on track to happen. Unfortunately, added Faigel, many small business owners would not apply for Paycheck Protection Program money because they were afraid of taking on debt or were suspicious of the Small Business Administration.

CLOSING REMARKS

To conclude the Boston workshop, Maria Flynn, president and chief executive officer of Jobs for the Future, and Robert Schwartz, professor emeritus of education at the Harvard Graduate School of Education and co-director of the National Pathways to Prosperity Network, reflected on some of the high points from the day's discussion, with an eye toward highlighting important policy implications for the labor market recovery in Boston and the nation. Then, Christine Abrams, president and chief executive officer of the Commonwealth Corporation, summarized the key priorities for greater Boston's workforce ecosystem and policy, business, education, and training institutions.

Both Flynn and Schwartz said they were optimistic, given what they had heard, that the recovery from the COVID-19 economic downturn would catalyze actions to overcome persistent economic development challenges that have been hard for the city to tackle during previous economic recoveries. In one sense, the national reckoning around racial equity and justice has brought more attention to this issue and enabled more direct discussions about addressing those inequities. In addition, the crisis is forcing more collaboration among organizations and learning from one another about how to do things differently. Schwartz added that his optimism stems from seeing people who have been involved in this type of work for decades, such as Rubin, Sullivan, and Polcaski, pivot so rapidly and adapt to the new challenges the pandemic created. For him, that signals that this may be the time to tackle these deeper problems in collaboration with new allies in a coalition of the willing that will bring its energy and passion around equity and social justice

An important lesson for Flynn was seeing that a coalition of the willing is the right way to go because it has more likelihood of having traction than if one of the stakeholders is taking the lead. In particular, she has been encouraged by the discussions that are bringing in all of the local institutions of higher education and the area's social service–led organizations to work on the problem of how to help individuals get on the first step of the career ladder and advance into a family-sustaining career and middle-class life. She does see a potential role for the federal government in incentivizing regional collaborations that go beyond local communities.

Community and Technical Colleges as Key Players

Schwartz commented on the importance of the nation's community and technical colleges. "If we are going to ever move toward having a real national workforce strategy, these 1,200 publicly funded community and technical colleges in my mind need to be at the center of it," said Schwartz. These institutions do not have anywhere near the funding to play the role they should be playing, particularly when it comes to retraining displaced workers. He noted that the Center for American Progress released a report in October 2020 documenting a $78 billion gap nationally between the revenues available to community colleges and the revenues available to four-year public institutions of higher education (Yuen, 2020), which in Massachusetts translates into a gap of approximately $8,000 in revenues per full-time student.

Flynn agreed with Schwartz about the importance of prioritizing funding community colleges to help them not only reach more students, but also be nimble and adjust to keep their offerings aligned with regional labor market demand. She also would like to see programs funded during the Obama administration, such as the Social Innovation Fund and the Workforce Innovation Fund, be restarted and to see more resources devoted to help students navigate the job market and career changes. Schwartz shared that he would also like to see resources directed to helping displaced workers find on ramps to new careers, whether by providing them with online learning opportunities or helping them with digital literacy, to take advantage of programs that can move them from disappearing or declining sectors into sectors with real opportunities to move into middle-skill jobs that provide a middle-skill wage.

In terms of drawing lessons from Boston's success and struggles that could be applied elsewhere, Flynn considered one of Boston's strengths to be the mix of legacy programs that have been in the community for years and the new programs that fit into the entrepreneurial system. Schwartz noted that one of Boston's continuing struggles has been its inability to develop a robust structure of introducing its public school students in a coherent manner to the world of work and careers. This work, he said, needs to start in the schools themselves. One lesson he has learned from examining strong vocational school systems around the world is that they successfully socialize young people into the world of work. The combination of work and learning should be better integrated in Boston.

Regarding the themes that Abrams said she heard throughout the day, the first was about rapid change being the new normal, which will require continual training for employees, the underemployed, and the unemployed. While moving fast will come with mistakes, the lessons learned from those mistakes will matter most, and using those lessons to adapt quickly in innovative ways will be a key to success moving forward.

Abrams' second takeaway was about how the pandemic has disproportionately affected people of color, low-income workers, women, and small businesses, and how it is further hollowing out the middle class. These most-affected populations have complex issues that go beyond a simple fix of finding a job. When it comes to increasing the number of people from those affected populations who can benefit from higher education, starting with community colleges, Abrams pointed to the importance of moving from diversity to social equity and challenging the status quo. Abrams also pointed out that the childcare model in the United States does not need to be revisited—rather, it needs to be re-engineered. The current system is broken, said Abrams, and "there needs to be a different way if we are to get women back into the workforce."

Abrams noted the importance of partnerships and alliances that the day's discussions highlighted. Partnerships that provide support and wraparound services and apprenticeships are about reinvesting into the economic infrastructure of a community. Toward that end, she thought the idea of using technology to provide wraparound services has merit.

Another theme of the day was re-engineering education with a focus on pathways and on ramps, particularly regarding skills and competencies rather than degrees. To advance that shift, employers need to be advocates for this type of training, which offer a more rapid way—than four-year degree programs—for people to get into high-paying STEMM jobs. Abrams reaffirmed that this is where community colleges can play a key role in a new workforce development strategy for the new normal. She noted that achieving true equity in the STEMM workforce requires rethinking and eliminating the bachelor's degree as a job requirement. Her organization, for example, is part of a collaborative partnership to redefine healthcare pathways that focus on competency-based learning and certificates and even using artificial intelligence to provide virtual hands-on training for some positions. Stackable credentialing and earn-while-you-learn will allow people to start working at better jobs sooner, as well as emphasize to workers the need to engage in lifelong learning to succeed.

4

Richmond, Virginia

Richmond, Virginia, is unique among the five cities on which the workshop series focused in that it is a capital city, with its large government workforce. As a result, there were several interesting state and local partnerships that this workshop, the third in the series, explored. At the same time, the city of Richmond, similar to the cities of Birmingham and Boston, had a much higher poverty rate, lower median wage, and more economic stress than the Richmond metropolitan area. As was the case for the previous two workshops, this one began with an overview of the economic situation in the city and surrounding metropolitan area, with the caveat that there are data through 2019, but up-to-the-minute economic data that can inform policy and practice are difficult to collect. This workshop also illustrated some of the ways in which Richmond's education and training programs have responded to the COVID-19 pandemic and how state and local policies and programs are meeting the challenges created by the pandemic.

UNDERSTANDING THE ECONOMIC AND WORKFORCE IMPACTS OF COVID-19 ON RICHMOND

In the workshop's first session, Stuart Andreason, director of the Center for Workforce and Economic Opportunity at the Federal Reserve Bank of Atlanta; Leonard Sledge, director of the Department of Economic Development for the city of Richmond; and Sonya Waddell, vice president and economist with the Regional and Community Analysis group at the

Federal Reserve Bank of Richmond, framed the workshop discussions by sharing economic, labor force, education, and demographic data for the Richmond metropolitan area and comparing the data to those from the state of Virginia. Alastair Fitzpayne, executive director of the Future of Work Initiative at the Aspen Institute, moderated a discussion with the panelists following their presentations.

Waddell noted that the Richmond metropolitan area has grown since the mid-1990s, although not as much as other similarly sized cities elsewhere in Virginia or in the South. Population growth in the city has also outpaced that of the metro area over the past decade as people move back to the city from its suburbs. In terms of employment growth and business formation, the city of Richmond has performed better than the United States as a whole. From her perspective, Richmond did not see the explosive rise in housing costs that northern Virginia, a suburban area adjacent to Washington, DC, saw prior to the COVID-19 pandemic, and the city had a strong base in professional business services and the types of jobs that could be done at home. It also has a large base in growing industries such as education, health services, trade, transportation, and utilities. Waddell explained that one thing that has hurt the city during the pandemic is that it had started to become a food and vacation destination, which led employment in the hospitality sector to more than double over the past few years and consequently crater, declining by 40 percent during the pandemic. At the time of the workshop, employment in the leisure and hospitality sector of Richmond's economy had recovered to about 80 percent of pre-COVID-19 pandemic levels.

With regard to poverty rates and wages, Andreason said that in 2019, the poverty rate and median wages in Richmond were 18.9 percent and $31,216, compared to 5.6 percent and $36,098 for the Richmond metro area, an imbalance that has existed for a considerable time. The demographics of the city and its metro area are significantly different, with the metro area being predominantly white and the city being more diverse with a nearly equal mix of Black and white residents (Figure 4-1). The difference in educational attainment between the city and metro area is relatively small, Andreason said, although the metro area does have an overall higher level of educational attainment (Figure 4-2). There is a high concentration of workers that have a high school degree or less in the city compared to the metropolitan area.

Unemployment in the Richmond metro area spiked around April 2020, hitting slightly over 11 percent, and women, adults under age 35, and Black residents were disproportionately affected. Across Virginia, while

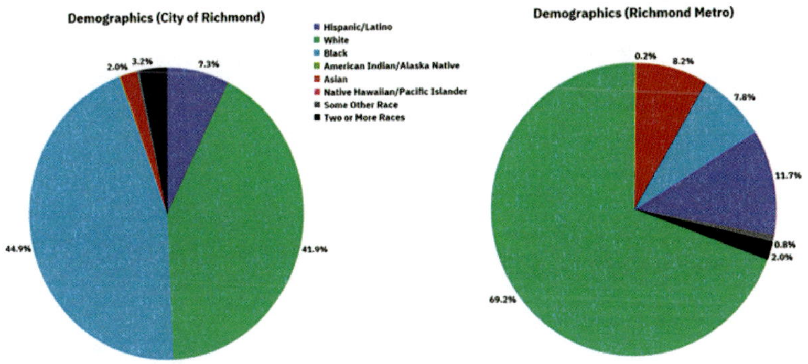

FIGURE 4-1 Demographics of the city of Richmond (left) and Richmond metropolitan area (right).
SOURCE: Andreason slides 3 and 4.

women made up 47.1 percent of the workforce, they constituted 55.7 percent of those who claimed unemployment benefits; people under age 35, who constitute 31.2 percent of the workforce, made up 42.3 percent of claimants; and Black people, who constitute 19 percent of the workforce, made up 38 percent of claimants (Figure 4-3).[1] During the week of September 26, 2020, some 47,000 more workers were on the new Pandemic Unemployment Assistance Program for gig workers or those otherwise unqualified for traditional unemployment than on traditional employment. In Virginia, many of the job losses have been in hospitality-related fields (Figure 4-4).[2] Andreason commented that the retail sector job losses accounted for a smaller share, at 6.7 percent of unemployment claims, than in the other two cities examined thus far.

Andreason explained that automation, as well as the COVID-19 pandemic, presents a risk of permanent job loss depending on a region's mix of jobs. The Richmond metro area's mix of jobs that are at risk of being lost permanently is similar to other portions of the south and mid-Atlantic regions, said Andreason, with those jobs at greatest risk being retail positions, cashiers, restaurant wait staff, cooks, and receptionists. The same

[1] Only state-level data were readily available, and Andreason noted that there are regional differences in the state.

[2] Unemployment claims by industry were only available by state, and not by city or metropolitan area.

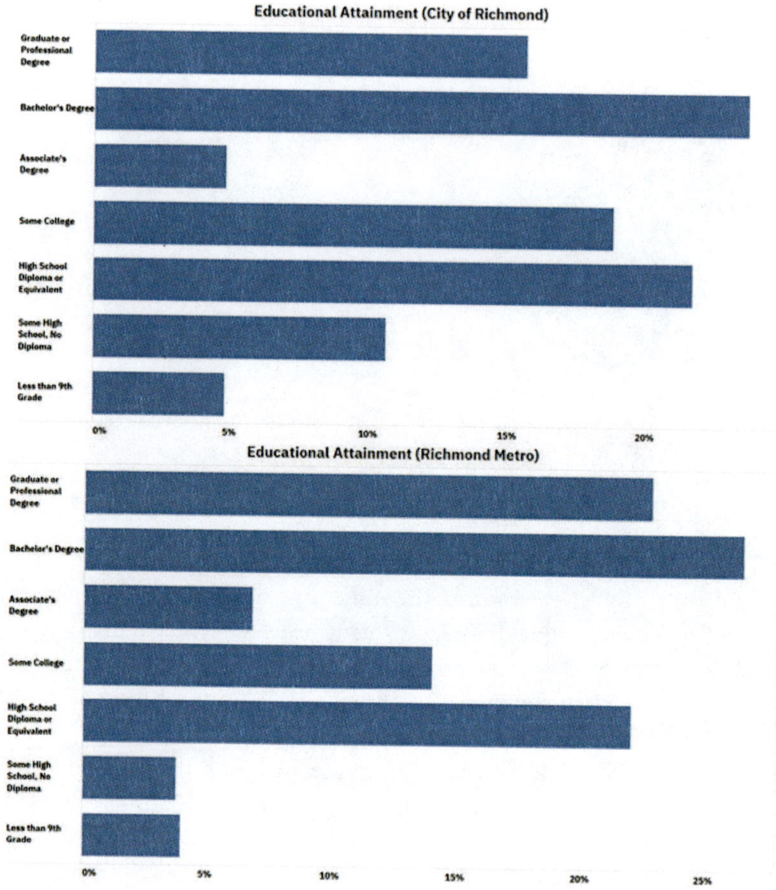

FIGURE 4-2 Educational attainment in the city of Richmond (top) and the Richmond metropolitan area (bottom).
SOURCE: Andreason slides 5 and 6.

can be said for the Richmond metro area's middle-skill, high-paying job mix in terms of its similarity to that of other southern and mid-Atlantic states. He noted that the Richmond metro area appears to be slightly more exposed than the rest of the state to permanent job loss and the percentage of workers who would be able to work at home.

Regarding internet connectivity, of key importance in allowing people to work or study at home, Andreason noted that the most recent data available show that across the Richmond metro area 85 percent of all households

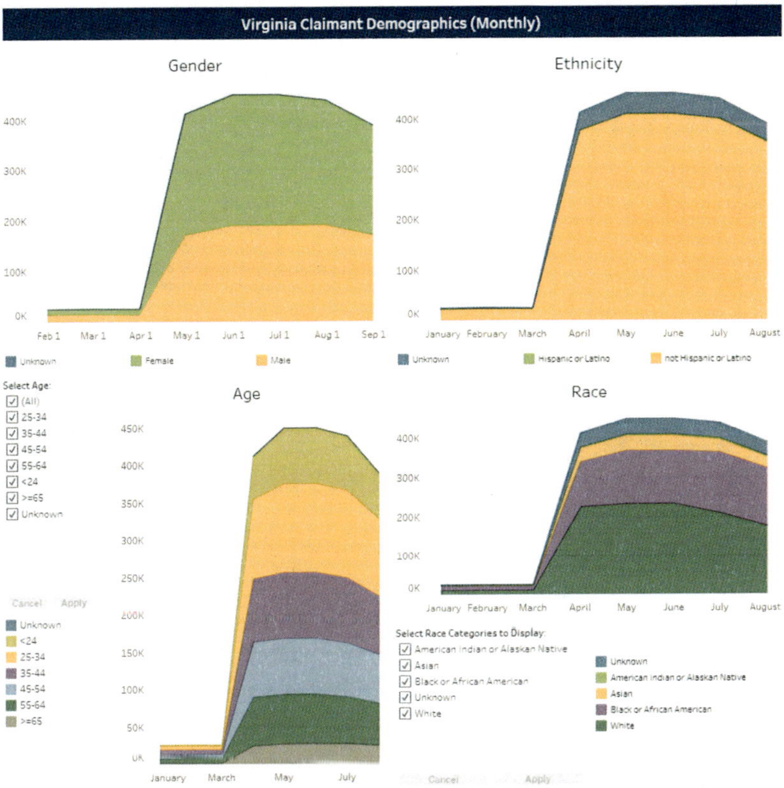

FIGURE 4-3 Unemployment claims for Virginia by age and race.
SOURCE Andreason slide 8.

have an internet subscription and 70 percent have a high-speed connection, compared to 82 percent and 68 percent, respectively, in the city proper. Slightly more than 11 percent of all households in the metro area, compared to 13 percent in the city, can only access the internet via their cellular plans. Andreason added that 39 percent of households in the metro area earn less than $20,000 annually and do not have internet access at home, compared to 35 percent of all households in the city. He also said that 78 percent of all households in the Richmond metro area have a computer at home, compared to 73 percent in the city. Some 9 percent of residents in the metro area rely on a cell phone for internet access, compared to 13 percent in the city, and 7.5 percent of residents in the metro area versus 9.4 percent in the city do not have a computing device at home.

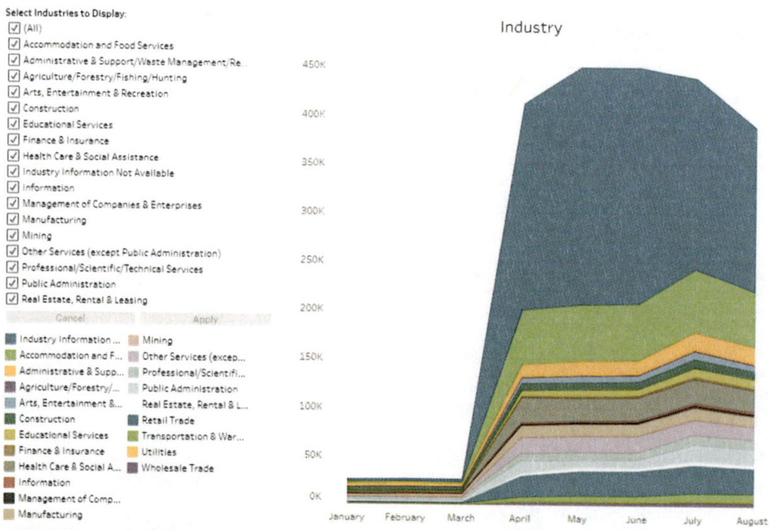

FIGURE 4-4 Unemployment claims for Virginia by industry.
SOURCE: Andreason slide 9.

Discussion

Sledge noted that the city of Richmond is focusing its efforts on and emphasizing equity and inclusion in its economic development activities so that Black and brown communities receive the benefits of these efforts. When asked whether there are any indications that the food service and hospitality workforce is exploring potential transitions to other industries, Sledge said that there is anecdotal evidence that individuals in the leisure and hospitality industry have been able to find employment in warehousing and logistics. He also noted that Richmond is now seeing a significant number of new restaurants and other hospitality and food service establishments opening, and there are employers looking for people even within the food service, hospitality, and restaurant industry. Waddell added that the pandemic-induced recession has been different from previous recessions in that the industries affected the most were those with a workforce less resilient to sudden changes. This suggests that retraining will be critical and that community colleges will need to play a large role in that retraining effort.

Andreason noted that he has been seeing people transitioning from hospitality to entry-level healthcare positions. He also pointed out that some sectors of hard-hit industries such as retail have been growing during the pandemic, for example, home improvement stores and grocery stores. In general, there has been less of an effect on science, technology, engineering, mathematics, and medicine (STEMM) fields, in part because there are ongoing increases in demand for STEMM employees, but also because many STEMM jobs can be done remotely.

One reason why Richmond has weathered the economic crisis caused by the COVID-19 pandemic a little better than other regions, said Sledge, is because of the presence of state government, local government, three research universities, and a major hospital system, as well as a major presence of banking and finance companies that can have their employees work remotely. He also noted the importance of the life sciences industry that is developing around the area's three major research universities and of making sure that the region's institutions of higher education are training a workforce capable of filling the jobs that industry is creating.

Fitzpayne commented that before attending the workshop he reviewed the most recent *Summary of Commentary on Current Economic Conditions* issued by the Federal Reserve Bank and saw that many businesses in the Richmond metropolitan area reported having difficulty filling positions.[3] Being somewhat surprised by this situation during a period of elevated unemployment, he asked Waddell if she has been hearing that from Richmond's businesses. Waddell replied that she had, but that unmet labor demand varies significantly by industry. For Richmond, the industries where the supply of labor is not keeping up with demand include construction, skilled trades, and manufacturing, with jobs in those areas requiring six months to a year of community college education. Sledge added that he has heard of companies in the area that are starting their own training programs to meet their needs for a trained and skilled workforce.

Andreason asked the other panelists whether Virginia's impending fiscal challenges—the same challenges that other states around the country are forecasting—will affect the local job market given that Richmond is the state capital. Sledge replied that the city had to make a $40 million adjustment to its budget once the pandemic hit, which led to a freeze on hiring and discretionary spending. However, local government is now filling some positions, although he was uncertain what the state government was planning in the future.

---

[3] Also known as the "Beige Book," this report is published eight times a year, and it summarizes current economic conditions in each Federal Reserve district. See https://www.federalreserve.gov/monetarypolicy/beige-book-default.htm.

RESPONSE OF RICHMOND'S EDUCATION AND TRAINING INFRASTRUCTURE TO THE COVID-19 CRISIS

The workshop's second panel discussed how the pandemic has affected workforce education and training programs for Richmond-area residents. The five panelists were Paula Pando, president of Reynolds Community College; Hakim Lewis, president of Virginia Union University; Peter Buckley, dean of the Virginia Commonwealth University School of Medicine; Brian Davis, executive director of the Capital Region Workforce Development Board; and Mark Barth, president and chief executive officer of Goodwill Central and Coastal Virginia. The session was moderated by Jane Oates, president of Working Nation, a nonprofit that works to create and amplify solutions for a changing economy, and former assistant secretary at the Employment and Training Administration of the Department of Labor.

Reynolds Community College, explained Pando, is one of 23 community colleges in Virginia and one of the four largest in the state, with four campuses in the Richmond area including one campus that houses its culinary hospitality and entrepreneurship programs. Community colleges, she noted, are a uniquely American invention, one created to fulfill a national belief that no person should not have access to quality postsecondary education or training, regardless of their life circumstances. She also commented that community colleges are not a destination but rather a pathway to further education or a job that leads to a living wage. At her institution that pathway includes partnerships with businesses and industries that need a skilled and diverse workforce, with K-12 education, and with various community partners that want to invest in Richmond's future.

Virginia Union University, explained Lucas, was started in 1865 in a jail in a town that was a cornerstone of the infrastructure of slavery. Through the perseverance of its first students—who were educated in that jail—and the community, the school now offers degree programs leading to bachelors, masters, and doctoral degrees for underprivileged individuals living in the Richmond area.

Buckley noted that Virginia Commonwealth University's focus during the pandemic has been to prevent job loss, continue to train healthcare professionals, and help the Richmond area safely open for business. Toward that end, the VCU Health System collaborated with local businesses to produce personal protective equipment, helped to educate the Hispanic and Latinx community about mask mandates, and launched an initiative to assist businesses and universities to reopen and provide guidance and testing strategies to prevent the spread of the novel coronavirus. VCU Health Systems also safely restarted in-person learning that is essential for licensure and credentialing of urgently needed healthcare professionals.

Connecting Job Seekers to Training Opportunities

Davis explained that Virginia Career Works is not a training provider but a connector that links job seekers to training opportunities and to employers in need of a trained and skilled workforce. In 2019, its three regional workforce centers served some 30,000 clients, placing 85 percent of them in jobs with an 80 percent retention rate and helping 75 percent attain the credentials needed for those jobs. He noted that at the time of the workshop, 8,300 employers had posted 40,674 jobs covering 716 different occupation types during the past 30 days. There were 466 unique certifications associated with those jobs, with six of the top 10 certifications pertaining to healthcare jobs. Projections for the year ahead showed that jobs with the highest projected growth were in healthcare, food preparation and food service, business and financial operations, and the construction trades. The challenge, he said, will be retraining individuals to fill many of those jobs given that 52 percent of people who lost jobs because of the pandemic have no postsecondary education.

Goodwill's mission, said Barth, is to help people help themselves through the power of work, and it does so by helping to address barriers to employment, whether those barriers are, for example, an intellectual or developmental disability, a criminal record, or a lack of transportation. In short, Goodwill's role is to get people on the pathway that Pando discussed. In 2019, Goodwill in Virginia placed 6,000 people in jobs.

Discussion

When asked how the COVID-19 pandemic has affected the way they do business, Barth replied that the biggest impact came from the closure of the Goodwill stores that fund its employment centers, costing the organization $9 million in revenues. Fortunately, Goodwill had received a grant in 2019 to pilot moving its offsite services, such as training, to locations it will share with other nonprofits in Richmond. Because that pilot program was successful in terms of meeting individuals where they are in the community, Goodwill is going to move to that model. He acknowledged, though, that this is a big shift for an $80 million organization that has been doing things the same way for 40 years.

Pando said that if there is anything that has been remotely positive about the pandemic, it is that higher education has accomplished tasks that she and others in the field thought would have been impossible, such as moving 900 class sections online in a matter of days. In her opinion, rethinking the entire model of postsecondary education will benefit the people who come to her institution. The pandemic also forced Reynolds Community College to lean into its values when it came to deciding what programs to cut in order to make room for expanding the programs that will train students for well-paying jobs and family-supporting careers.

At Lucas's institution, the pandemic has accelerated the processes that required thinking of education as more of an entrepreneurial business. In fact, he now thinks of Virginia Union University as an educational technology startup that is building a business incubator and accelerator for people of color. He also noted that the realities of the Black Lives Matter movement on the heels of the COVID-19 pandemic and its disproportionate effect on Black and Latinx individuals revealed that access is the primary barrier to workforce development, training, and eventual employment. One aspect of his institution's response has been to create new certificate programs and master's degrees, as well as an associate degree, as a way of recruiting a diverse student population.

Davis said the Capital Region Workforce Development Board has not gone through the same seismic shift that educational institutions have had to endure, but one shift it did have to make was to learn how to remain eligible for federal employment and training program funds in a virtual format. This involved installing new systems for verifying signatures and program eligibility and transmitting personally identifiable information electronically. He said his organization has done a good job staying in touch with current clients, but new client enrollment has fallen since the pandemic began.

One of the most radical changes for the VCU Health System was going from approximately 80 telehealth visits in any given week to nearly 6,500 once the pandemic started. This change did not merely affect the delivery of care, but also forced Buckley's institution to determine how to best train the workforce to deliver care in this new virtual environment. He acknowledged how important the community's outpouring of support for VCU Health System's workforce has been in terms of the morale and loyalty of that workforce as well as in increasing the community's interest in careers in healthcare and science.

Regarding how their institutions are pushing for more universal STEMM education, particularly for populations that are underrepresented in the current STEMM workforce, Pando said that her institution's laser focus is on people living in communities of color—that is, on who enrolls at Reynolds Community College—and getting to those individuals and communities early to make them very aware that there are careers in STEMM that pay well, not just in social work or human services. She then challenged her business and industry partners to do a better job of providing role-modeling types of relationships to students in middle school, when they are starting to consider what careers they might like to pursue. Many of the students her institution serves have never heard of a welder, seen an electrician, known what a lab technician does, or been aware of the difference between a registered nurse and a certified nursing assistant. In addition, representation matters when it comes to role models, she said, because students relate more when the role model looks like them or comes from the same type of background.

Van Freeman, director of the Aerospace Industries Association's future workforce development and strategy initiatives, asked the panelists what the colleges and universities in Virginia are doing to alleviate the burden on students of color who are trying to enter STEMM fields, given that they are likely to have more student debt that their white counterparts. Lucas replied that Virginia Union University's approach has been to build degree programs that lead to well-paying STEMM jobs and to procure opportunities for them to get hands-on experiences in those fields, allowing students to conduct research in a laboratory and earn while they learn.

Buckley said that the VCU Health System has been on a long-term journey to match its trainees' ethnic and social distribution with that of the state's population, because people have the best health outcomes when they are cared for by somebody of similar racial and ethnic background. He noted, too, that the outpouring of financial support his institution has received during the pandemic will allow it to support the debt of its students of color, which he hopes will convince more of those students to practice in the community once they finish their training.

INNOVATIVE WORKFORCE SOLUTIONS: HOW STATE AND LOCAL POLICY AND PROGRAMS ARE MEETING THE COVID-19 CHALLENGE

The day's third panel examined post-COVID policy responses and investments in workforce training programs, including the effect of the pandemic on workforce alignment and learner access. The five panel members were Megan Healy, chief workforce advisor to Virginia Governor Ralph Northam; Valaryee Mitchell, director of the city of Richmond's Office of Community Wealth Building; Caren Merrick, chief executive officer of the Virginia Ready Initiative, a newly established nonprofit organization formed in response to the economic hardships created by COVID-19 that works with local partners to reskill people for in-demand jobs in high-growth sectors; Elizabeth Creamer, vice president of workforce development and credential attainment at the Community College Workforce Alliance, a community college partnership that provides non-credit training, custom-designed instruction, consulting, skills assessments, and educational programs; and Nat Marshall, chair of the Virginia Board of Workforce Development. Anna Payne Fife, associate director for outreach and engagement for the Aspen Institute Future of Work Initiative, moderated the discussion among the panelists after their short presentations.

Credentials to Move People into Recession-resistant Industries

Healy noted that the state has done relatively well avoiding layoffs of employees at the high end of the salary range, many of whom were able to continue teleworking, so the challenge going forward is moving people who worked in lower-wage jobs in restaurants and retail into recession-resistant industries such as those that are STEMM-based. Three important pieces of addressing that challenge will be providing wrap-around supports for students and workers, including childcare and transportation; using an equity lens when launching new programs; and leveraging technology to make worker training accessible and affordable for every resident who needs it. The equity lens is key, said Healy, to avoid what happened during the recovery from the 2008 recession, when it took people of color longer to find jobs and recover economically.

Creamer's organization, the Community College Workforce Alliance, serves more than 9,000 regional residents per year, providing employees for over 200 businesses and organizations in the region. To get residents of greater Richmond back to work, her organization works through industry sector strategies that focus on specific industries and then uses what she called FastForward workforce credentials, apprenticeship programs, and pre-employment training.[4] It also does contract training for business

[4] Additional information about FastForward workforce credentials is available at https://www.fastforwardva.org/.

and has programs to upskill incumbent workers, and it recently joined a regional workforce development coalition called Network to Work.

Merrick explained that after studying the local economy in detail and interviewing 35 of Virginia's top companies to see where the job openings were concentrated, the Virginia Ready Initiative partnered with the Virginia community college system to recruit people into the FastForward credentialing program to train people for jobs in healthcare, manufacturing, the skilled trades, and information technology and cyber technology. The most popular credential so far has been for billing and coding, which opens the door for great jobs in healthcare, she said. The initiative is recruiting people who are unemployed or underemployed or whose employment has been disrupted by the COVID-19 pandemic to take one of 30 credential programs. Those who take advantage of the program and receive a credential are awarded $1,000 as an incentive and become eligible to take advantage of jobs posted by any of the 23 Virginia businesses (and counting) that have partnered with this program and that help with curriculum updates at the state's community colleges. Merrick noted that some of the business partners had never worked with a community college and said that the goal is to have 2,000 people finish this program by the end of 2020 and to have supported a minimum of 15,000 people over the next two years.

Discussion

When asked whether Richmond was deploying any creative solutions to address the broader challenges and obstacles, such as lack of childcare, transportation, and internet access, that affect workers' ability to secure employment and stay employed, Mitchell described an initiative begun by the Office of Community Wealth Building called the ambassadors program. Through this program her office recruits community members who act as two-way conduits of information, sharing information with the community and providing information about community needs to her office and to other partners. One surprising finding by the ambassadors was that many community members were not aware of food programs and COVID-19 testing sites, even though the city had a significant social media effort and created websites to publicize these programs. The problem was that too many of the people the programs were intended to support did not have internet access or a smartphone and were therefore essentially blind to the social media effort. Within a week of discovering this problem, the ambassadors were blanketing the community with paper notices about these programs. The city, said Mitchell, is also working to increase the number of cell phone towers near Richmond's public housing communities as a means of addressing this access problem.

Regarding childcare, the city has partnered with the YMCA, which reduced its price for childcare so that low-paid essential workers could afford it. The city is also working with other childcare providers to see how they can work together to provide more affordable and safe childcare for workers who cannot afford to pay $150 per week. Her office is now going into the community to make sure that school-age children are, in fact, engaging in virtual learning, and if not, to look for solutions to whatever barriers are keeping them from participating in online schooling.

When asked about the policies the state has implemented since the COVID-19 pandemic began, Healy said one of the first actions the governor's office took was to create one website to serve as an access point for all available data on regional workforces. The state has also developed an online referral tool that individuals can use to enter their information and get referrals to jobs or social services, and has secured federal funds to address those social needs. The governor allocated $5 million for community colleges to "flip" their workforce training pathways so that job training, rather than coming later in a student's course of study, starts on day one. Someone who wants to be a welder, for example, will start with a welding class, not English or math.

When asked whether she is finding it harder to develop partnerships during the COVID-19 pandemic, Creamer said that there has been renewed energy around establishing partnerships thanks to work that had been done before the pandemic. The Virginia Ready Initiative, for example, formed quickly once the pandemic began. She noted, too, that employers in the region are more conscious of the need to work in partnership with community organizations to address systemic racism and fundamentally change the outcomes at all levels of education for students of color. The result of that increased awareness is a great deal of effort around addressing the digital divide and increasing access to childcare and transportation, with the goal of better preparing students for family-supporting jobs. Creamer added that there are now 39 different groups—public, private, and philanthropic—coming together to lend their talents in a coordinated way to address local inequities. In short, there has been "an explosion of interest in partnerships," as well as a commitment among employers to guarantee jobs in manufacturing, logistics, and healthcare for students coming out of the many programs that were discussed during the day's workshop sessions.

Marshall commented that industry partners need to do more than just have jobs available, and should work with community college and economic development partners to train workers to fill those jobs. In his opinion, employers need to go into schools and talk about the kinds of jobs that students can get with some postsecondary education, and that employers should do so regularly and repeatedly. As an example, he has gone into middle and elementary schools talking about the opportunities for machinists, who can start with wages in the $30 to $45 an hour range, with time-and-a-half pay on Saturday and double time on Sunday. When the students think about that possibility compared to the minimum wage they could make working at a fast-food restaurant, they get excited about work possibilities.

In terms of actions the panelists would like to see from the federal government to help more Virginians return to good jobs, Merrick said she would like to see the federal government revisit the types of security clearances needed to work for a federal contractor or for the federal government. Those security clearances can be a barrier, and she suspects that some of positions do not need one as high as is currently required. Creamer said she would like to see the federal government make it easier for working adults to get ahead by accessing federal financial aid or workforce training at community colleges and universities, particularly for part-time students. Marshall wanted to have the federal government continue to support increasing broadband internet access for everyone in the state, and Mitchell said she would like to see the federal government look at the "cliff effect" for benefits. The cliff effect occurs when people who are receiving benefits start losing them rapidly when they do get a job, which often leads to people struggling to pay for the childcare and transportation they then need to keep that job.

Healy agreed with all of the suggestions and stressed the need to support both training and wrap-around services. As she noted, life is expensive, and navigating those expenses for someone embarking on a new career path can be overwhelming. Any support that students can receive will improve their chances of succeeding.

CLOSING REMARKS

To conclude the workshop, Andreason and Josh Carpenter, director of economic development for the city of Birmingham, provided summaries of the key themes, challenges, and policy opportunities they heard over the course of the day. Andreason noted the great opportunities for cross-sector collaboration and the need for federal action. He was struck by the need to address credit checks, which can also be a barrier to employment for people who, for example, lost their homes during the pandemic. Employers, he said, need to decide whether credit checks, in the current environment, are a relevant criterion.

Carpenter was impressed by the resonance he heard among the different panel members regarding the need to shift institutional focus and work in partnership to address the immediate and long-term economic issues highlighted by the COVID-19 pandemic. In particular, he thought that the Richmond workforce system was well connected across the community, thanks to leadership that was willing to be adaptive and form partnerships. Community college leaders, for example, have been receptive to changing their offerings, even overhauling the design of their courses, in order to better meet the needs of local employers. Based on what he heard over the course of the day, Carpenter said that rethinking the way postsecondary education is funded is essential if the goal is to help people who have been disadvantaged get the training they need to secure better-paying jobs. He also highlighted the importance of increasing students' exposure to potential jobs early on and going into schools to talk to elementary and middle school students about career opportunities with relatively better-paying jobs.

5

Riverside, California

The fourth workshop in the virtual workshop series focused on creating equity within the workforce pipeline and connecting it to the green economy and the high technology, high-growth jobs of tomorrow in Riverside, California. This workshop also higlighted the importance of Riverside and the surrounding region's logistics industry and its role in addressing workforce issues in the post-COVID-19 pandemic labor market. In introductory remarks to the workshop, Eloy Ortiz Oakley, chancellor of the California Community Colleges, commented that people are adapting to life during the COVID-19 pandemic and are learning how to work, shop, and interact with one another in innovative ways. Those changes create opportunities, and one goal of the day's discussions, he said, would be to discuss today how these these changes can be beneficial to those seeking new jobs or career paths in and around the greater Riverside region.

UNDERSTANDING THE ECONOMIC AND WORKFORCE IMPACTS OF COVID-19 ON RIVERSIDE

Riverside is unique among the cities selected for the workshop series in that the city of Riverside has a lower poverty rate than the surround metropolitan area—11.5 percent versus 13.1 percent—although it does have a lower median wage than the metropolitan area as a whole, at $32,431 versus $35,713. Riverside differs, too, as do other cities in the western United States, in that its counties are significantly larger geographically, which plays out in terms of demographic data and the ability of organizations and government to act on new programs and policies. Riverside was the first city examined by the workshop series with a predominantly Hispanic or Latinx population (Figure 5-1). Both the city and metro area are similar in terms of educational attainment to the other cities in the workshop series (Figure 5-2).

Andreason described data showing the unequal impacts of the COVID-19 pandemic. Unemployment in Riverside and the metro area, he explained, was quite low prior to the COVID-19 pandemic—4.7 percent for the city and 4 percent for the Riverside-San Bernadino-Ontario metropolitan area—but it had risen and remained significantly higher at the time of the workshop. The unemployment numbers for the city and metro area were more closely aligned than for other cities, and Andreason said he suspected that something interesting is happening in the Riverside area. He noted that, as in the other cities, women, adults under age 35, and Blacks have fared disproportionately worse than the local workforces as a whole in terms of unemployment. The Hispanic/Latinx population, however, was slightly less affected, accounting for 37.2 percent of the workforce but only 36.1 percent of those who claimed unemployment as of September 2020 (Figure 5-3). He noted as well that during the week of September 26, 2020, nearly 700,000 more workers were on the new Pandemic Unemployment Assistance Program for gig workers and the self-employed than were on traditional unemployment.

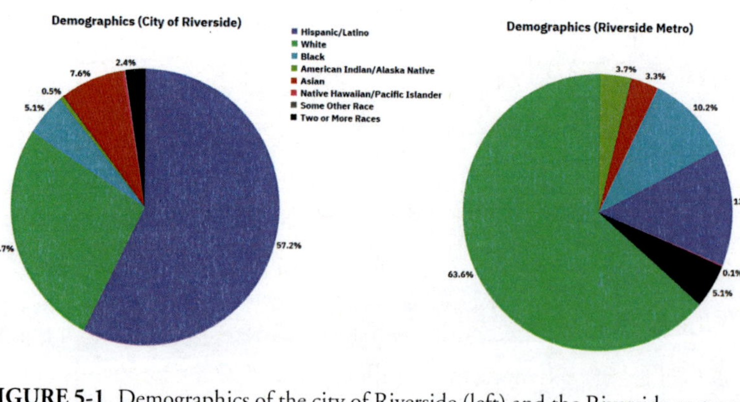

FIGURE 5-1 Demographics of the city of Riverside (left) and the Riverside metropolitan area (right).
SOURCE: Andreason slides 3 and 4.

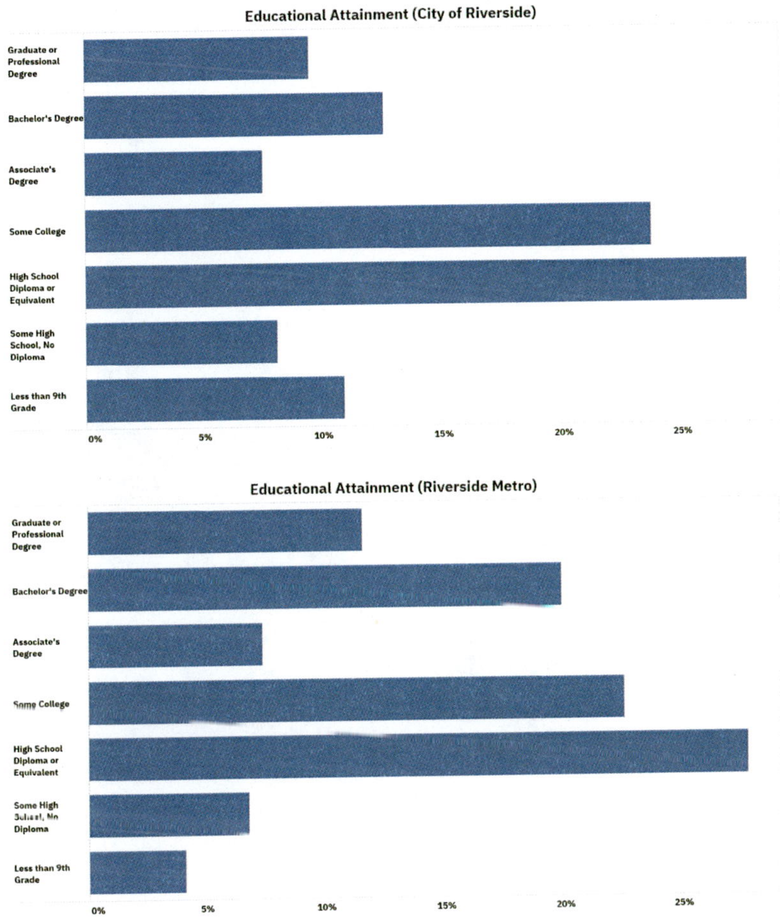

FIGURE 5-2 Educational attainment in Riverside city (top) and Riverside metropolitan area (bottom).
SOURCE: Andreason slides 5 and 6.

In California broadly, unemployment claims by industry also illustrate an unequal effect of the pandemic, with retail trade, accomodation and food services, and healthcare showing the biggest increases in the percentage of the state's unemployed workforce (Figure 5-4), and unemployment in the retail sector surpassing that of much of the rest of the country.[1] In total,

[1] Unemployment claims by industry were only available by state, and not by city or metropolitan area.

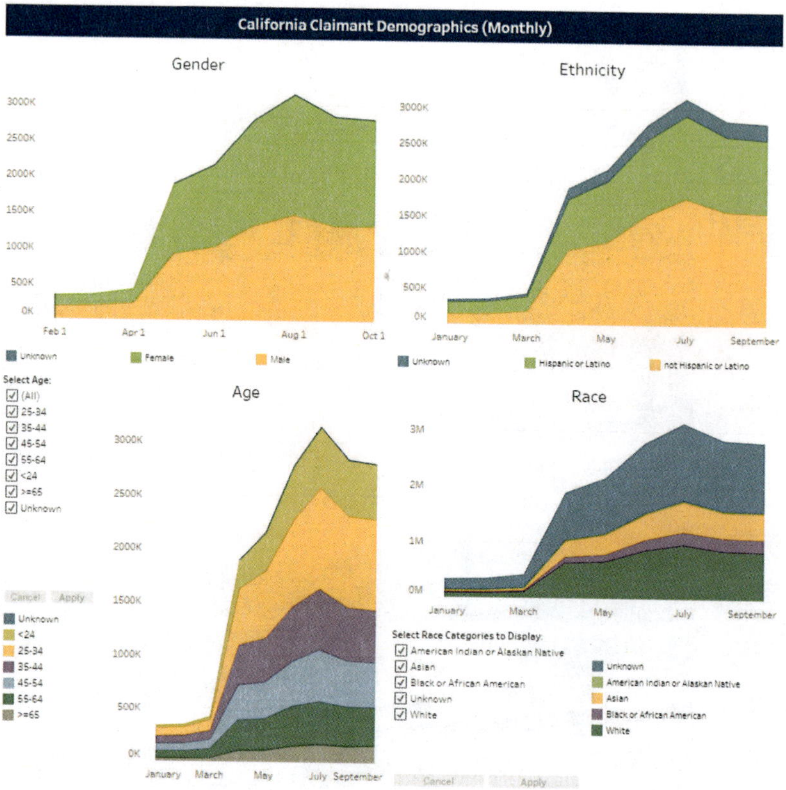

FIGURE 5-3 Unemployment claims for California by age and race.
SOURCE Andreason slide 8.

unemployment in the Riverside metropolitan area is above average but similar to other inland metro areas in California, and total employment is more at risk of permanent job loss as a result of the pandemic than many areas in California and other western states. The Congressional Budget Office estimates that 61 percent of the Riverside metro area's industry employment demand is expected to be affected compared with 49.6 percent statewide, while work-at-home estimates suggest that only 24.7 percent of the region's workers would be able to work at home compared with 28.4 percent statewide. Andreason commented that at-risk opportunity jobs—those middle-skill positions that pay above the median regional wage and do not require a bachelor's degree—are concentrated in the retail and hospitality industries, as well as in some construction sectors.

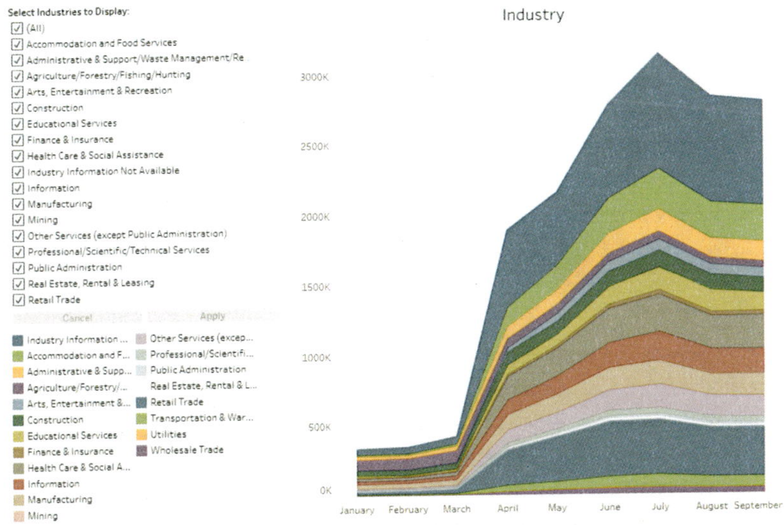

FIGURE 5-4 Unemployment claims for California by industry.
SOURCE: Andreason slide 9.

Almost 90 percent of the Riverside metro area's households have an internet subscription, three-quarters of the households have a high-speed connection, and just over 10 percent access the internet solely through cellular data plans, which Andreason said is better than in many metropolitan areas. Under 10 percent of all households in the area have no internet subscription, which includes nearly 28 percent of households earning less than $20,000 a year. In terms of ownership of computing devices, just over 80 percent of all households have a desktop or laptop computer and less than 10 percent have only a smartphone.

GREATER RIVERSIDE ECONOMIC OUTLOOK PRIOR TO AND DURING COVID-19

The first panel discussed investments in workforce training programs prior to the COVID-19 pandemic and the affect of the pandemic on their workforce alignment and learner access. This panel also considered inequities in the workforce and how to address them. The three panelists were Kim Wilcox, chancellor of the University of California, Riverside (UCR); Toni Symonds, chief consultant to the California State Assembly's Committee on Jobs, Economic Development, and the Economy; and Wolde-Ab Isaac, chancellor of the Riverside Community College District (RCCD). Eloy Ortiz Oakley, Chancellor of California Community Colleges, moderated the session.

Working Together to Address Inequities

Wilcox proposed two topics as important for the day's discussions. The first was the extent to which different organizations and institutions work together in the region, which he said was a real strength of the Riverside area. The second was the work going on in the "inland empire," as the region is known colloquially, to address inequities in workforce culture. To that end, he noted that UCR was voted by *U.S. News and World Report* as the leading university in the country for social mobility; it ranks first in the nation in Hispanic enrollment among selective U.S. public universities according to the Urban Institute and third in the nation for the number of Hispanic students completing science, technology, engineering, mathematics, and medicine (STEMM) degrees according to the National Science Foundation; and more Pell grants have been received by UCR students than in the entire Ivy League. All told, 57 percent of UCR's undergraduates are first-generation college students, and it is among the best in the United States, according to Education Trust, in terms of graduation success for Latinx and Black students.

Riverside County, explained Wilcox, is the 10th most populous in the country and ranks fifth nationally in population growth. At the same time, it has the greatest shortage of primary care and specialist physicians in California, which is why UCR's seven-year-old medical school is "laser focused" on equity and the diversity of its medical school classes and the ways in which it engages with its racially and ethnically diverse community.

Wilcox described four collaborations that he believes are defining for UCR:

- Growing Inland Achievement, a collaboration among educational institutions and the private sector focused on increasing the number of credentialed individuals in that region.
- Inland Economic Growth Opportunity group, another collaboration among educational and private sector groups focused on developing good jobs for the entire region.
- Riverside County Educational Cooperative, a group focused on improving the college-going culture in the region.

- University Innovation Alliance, a national coalition of public research universities committed to increasing the number and diversity of college graduates in the United States.

These collaborations, the first three of which also involve RCCD, have overlapping goals focused on how to prepare all of the citizens in the Riverside region for the jobs of today and the future. Wilcox noted that most of the institutions involved in these collaborations are under-resourced. UCR, for example, receives fewer state dollars per person than any other campus in the University of California system, evidence of the systematic, historic differences in how resources are invested in different regions of the country that still play out today.

Helping Educationally and Economically Challenged Populations

RCCD, said Isaac, serves more than 60,000 students annually, with Hispanic students accounting for almost two-thirds of that enrollment. Nearly three-quarters of the students attend part-time, more than 40 percent are first-generation, more than half are women, and 13 percent are from what he called special populations that include students with a disability, veterans, children raised in the foster care system, and adults who are or have been incarcerated. He noted that over 70 percent are of low socioeconomic status and therefore receive free tuition. During the pandemic, RCCD had 26 laboratories open in which some 2,400 students were completing their practical studies in a highly secured and safe working environment.

RCCD and the other educational institutions in the region are all deeply aware that the population they serve is both educationally and economically challenged, with one of the nation's lowest rates of college attendance and completion. This recognition, said Isaac, is what has driven the collective and collaborative approach that the regions' institutions are taking and that has led to strong, durable partnerships spanning the entire educational landscape, from kindergarten through the baccalaureate degree, and includes the region's business and industry sectors. He also noted that RCCD's workforce development program is designed to address equity, with a focus on students with a disability, those coming from economically disadvantaged families, single parents, people experiencing homelessness, youth in the foster parent system, and people who are displaced.

In 2019, Isaac's institution's graduation rate increased by 27 percent even though it has the fewest faculty positions per full-time-equivalent student in the state's entire educational system and the smallest total amount of financial aid provided to its students. "It is ironic that while we are the champions of equity, we are also the victims of inequity," said Isaacs, noting the community colleges in California produce nearly 70 percent of all middle-skilled workers. He pointed out that by 2025, the Riverside region's secondary schools and community colleges will produce more than 80,000 students who qualify for admission to a postsecondary educational institution, requiring a significant investment in the expansion of the region's public institutions if the region is to achieve its economic development goals.

Boosting the Demand for Jobs

Symonds noted that California entered the COVID-19 pandemic with historically low unemployment, although in certain areas of the state, including the "inland empire," some groups of people have not benefited equitably from the economic expansion following the recession of 2008. As an example, she cited figures showing that while overall poverty in the state was trending downward, some 6.7 million Californians—including 19 percent of Black and 15 percent of Latinx individuals—lived at or below the federal poverty threshold, compared to 8.3 percent of white residents of the state. In Riverside County, 14.4 percent of Black, 12.8 percent of Latinx, and 17.9 percent of indigenous individuals, compared with 9.9 percent of whites, lived at or below the federal poverty threshold.

The state legislative committee she works for has been focusing more on the demand side—that is, employers—rather than directly on workforce development during the COVID-19 pandemic. A key aspect of this has been supporting small businesses, which historically have led economies out of recession. In California, net new job growth among innovation-based businesses was the primary engine powering the state's economic growth prior to the pandemic, and the challenge, said Symonds, is helping those small businesses pivot to meet current and future market demand. Doing that requires understanding the new skills that workers will need to compete for positions in the new, post-pandemic environment.

Symonds explained that while the state budget has undergone a dramatic reduction during the pandemic, it was still able to allocate $300 million to support K-12 technology programs and maintain funding for workforce development and adult education programs at the state's community colleges. The legislature also provided $5.8 million to support services related to career pathways, such as childcare programs; $20 million to support small businesses; $10 million for social entrepreneurs to support an economic development initiative to assist economically disadvantaged groups, particularly those with limited English proficiency; and $37 million for an adult re-entry grant program.

The state's community colleges, said Symonds, have partnered with a statewide program called the California Employment Training Panel to start new health and safety courses for frontline workers and supervisors and to update virtual employer training courses. The program's current emphasis is on short-term certificate programs that can get people into jobs immediately. Symonds also highlighted an initiative that the Western Riverside Regional Council of Government has been working on, called Innovation 2030, that is trying to break down the silos that separate resources for entrepreneurship from workforce development efforts.

Discussion

Regarding federal assistance provided during the COVID-19 pandemic, Symonds noted that the federal government's supplemental unemployment insurance program funds are fundamental if the goal is to help small and large businesses alike survive until the pandemic is controlled and businesses are able to operate without restrictions. She would also like to see some federal support for expanding broadband internet access and providing computers for people who are economically disadvantaged.

Isaac added that investments in the state's community colleges would be valuable. He and his colleagues have been working to shorten the time to graduation by 50 percent, which means reducing the cost per degree by half and doubling the available space for training. Any investment by the state or federal government, therefore, would be small compared to the resulting value: what community colleges would produce in terms of a middle-skilled workforce. The bottom line is that at the state and federal levels, the significance of community colleges has not been appreciated properly, said Isaac. Wilcox seconded Isaac's wish for more funds to support community colleges and added that the federal government needs to help shore up California's finances, which will ultimately help the state's community colleges and universities. At the same time, he wants to see the state and its educational system have the gumption to reallocate existing money to support programs that address inequities.

REGIONAL ECOSYSTEMS AND OPPORTUNITIES

The workshop's second panel focused on emerging workforce training programs in the Riverside region, including apprenticeships and credentialing programs. The four panelists were Leslie Trainor, deputy director of workforce development for the County of Riverside; Noel Hacegaba, deputy executive director of the Port of Long Beach; Rex Beck, professor of business administration at Norco College; and Chris Cagle, regional affairs manager for the South Bay Workforce Investment Board. Sid Voorakkara, senior vice president of Strategies 360, moderated the session.

Building a Healthy Workforce Ecosystem

Riverside County, explained Trainor, has 28 cities—Riverside, Palm Springs, and Coachella among them—and various incorporated areas encompassing 7,303 square miles, which is larger than four U.S. states. Its population is close to 2.5 million people, making it the 10th most populous county in the nation. The scale and size of the county, said Trainor, makes creating policy and administering workforce development efforts tricky. She and her colleagues see workforce development as an ecosystem in which workers and employers are interdependent entities. For the workforce system to operate optimally, both workers and employers must be "healthy," which means that her office looks to enable a skilled and prepared workforce while simultaneously ensuring that the county's employers are strong and resilient.

The services Trainor's department runs are directed at three groups: job seekers, youth aged 16 to 24, and businesses. For job seekers, programs include career coaching and job-readiness workshops that include providing help with preparing resumes and conducting mock interviews. Her office can also subsidize vocational training up to $8,000 per person, and it runs three American Job Centers funded by the Department of Labor. For the county's youth, her department offers career exploration opportunities, skills training, and educational development programs. It also runs six youth centers around the county. Youth programs, said Trainor, are funded by the federal Workforce Innovation and Opportunity Act.

For businesses, Trainor's office provides a variety of support services, including on-the-job training for new and existing employees that can include subsidized pay of up to 50 percent of a new employee's salary during the training period. Other services include recruitment, application screening, hosting job fairs, providing assistance with human resources issues, and helping employees affected by the COVID-19 pandemic rapidly shift to new positions or apply for unemployment benefits.

Trainor noted that the county targets industries that are projected to grow countywide in the coming years and pay a living wage without requiring a college degree, but perhaps requiring a training program of 18 months or less. These industries include healthcare, clean energy, manufacturing, construction, logistics and transportation, retail, information security, and administrative support. The distribution of industries varies across the county; for example, the eastern part of the county will need employees for crop production, while the western part will need employees in architectural and engineering services, advertising and public relations, legal services, insurance-related services, and wholesale services for various materials.

Improving Supply Chains to Benefit Workforce Development

The Port of Long Beach, which lies just west of Riverside County, is the second busiest U.S. port, generating about 51,000 jobs in the city of Long Beach, nearly 600,000 jobs across southern California, and some $5 billion annually in tax revenues. Hacegaba noted that the nation's ports generate $5.4 trillion in total economic activity, or 26 percent of the nation's economy. Workforce development, said Hacegaba, has always been important to his industry.

Hacegaba noted that the COVID-19 pandemic has triggered some longer-lasting changes that will affect workforce development. For example, the doubling of e-commerce has ramifications for last-mile logistics and the type and number of jobs required to support that change. The peaks and changes in cargo demand has heightened the need to leverage technology to help inject visibility and predictability into the supply chain, which will generate more jobs requiring technology skills and opportunities for employees to add additional skills to their resumes. In addition, the wave of baby boomer retirements among port workers will require training a new workforce in the relatively near future.

He also expects that as the nation comes out of the pandemic, it will need new operating paradigms aimed at making supply chains more efficient, resilient, predictable, and reliable. At the same time, he said, there will continue to be a growing emphasis on emissions reductions, which require new technologies and employees with the skills to operate and maintain those technologies.

Supporting the Logistics Industry

While the Port of Long Beach is not in Riverside County, it is close enough to make the region around Riverside one of the largest logistics hubs in the United States, explained Beck. He noted that Norco College, one of the region's community colleges, offers apprenticeships, internships, and short-term non-credit classes and certificate programs to support that industry. In fact, Norco College offers a unique certificate program and full degree program in logistics management of the type that is normally taught as part of a master's degree program. Over the past 18 years, this program has awarded over 330 certificates and degrees in logistics management. The lesson here, he said, is if there is an industry that at first glance seems as if it requires graduate degrees or upper-division coursework, consider how that might be done at the community college level instead. The one challenge with this is that upper division courses taught at a community college may not transfer to a four-year institution.

Beck said that given his institution's focus on getting people back into the workforce and providing additional training for those who need new skills, the college only offers courses in the evening or online to make it easier for workers to attend classes. Norco College is also developing something Beck called a degree completion program that will enable someone with training in diesel mechanics to gain course credits based on their experience and the courses they took to get their certificate and then complete the courses they need to then enter an engineering program at a four-year institution.

Creating Opportunities for On-the-Job Training

Cagle discussed pre-apprenticeship and apprenticeship programs that the South Bay Workforce Investment Board is running in neighboring Los Angeles County. The program is driven by employers' workforce needs, with a focus on the aerospace and bioscience industries. What makes the Aero-Flex and Bio-Flex programs unique, he explained, is that each employer can tailor coursework to fit the specific occupations for which it needs employees. The pre-apprenticeship program lasts 60 to 90 days and includes three tracks: work-readiness training, industry-specific training, and work-based learning or on-the-job training that includes a small stipend. The latter two tracks can be done virtually.

The Bio-Flex program, said Cagle, provides training for production and manufacturing technician, maintenance technician, quality control analyst, and, most recently, lyophilization maintenance technician. The Aero-Flex program trains people for jobs as an aerospace engineer, computer numerical control machine operator, industrial machine system technician, industrial engineering technician, and aerospace electronics technician.

Cagle described how his organization, with employer direction, developed Aerospace 101, a free course for anyone who wants to get an idea about what working in the aerospace industry would be like. He and his colleagues are also developing a youth apprenticeship program to allow high school seniors to get a part-time job, continue in a full apprenticeship program after graduation, and attend community college to get a certificate or two-year degree in an applicable field. Together with local employers, the investment board uses apprenticeships as a strategy for workforce and pipeline development that can take a high school student and, over the course of four years, guide them through an apprenticeship program, community college degree, and bachelor's degree, with the student being employed by a company throughout the process. This type of hybrid experience not only provides technical training, but enables companies to give students experience in the business side of the company's operations. The Department of Labor recently awarded Cagle's organization and partner West LA College with funds to put 5,000 workers into pre-apprenticeship and apprenticeship roles in advanced manufacturing for the aerospace and bioscience industries. Cagle's organization has secured funding to provide participants in its programs with a laptop computer, and the organization is working with the South Bay Council of Governments to build a new broadband network that will connect all 15 cities in the area and offer much less expensive broadband connections to the public.

HIGHER EDUCATION'S ROLE IN ECONOMIC MOBILITY

The day's final session explored current and future efforts by institutions of higher education to support local workforce development, with particular attention to increasing the diversity of the STEMM workforce pipeline. This session also discussed technology transfer and the growth and expansion of entrepreneurship. The three panelists were Jeannie Kim, vice chancellor of educational services for RCCD; Rodolfo Torres, vice chancellor for research and economic development at UCR; and Byron Ford, associate dean for pre-clerkship medical education at the UCR School of Medicine. Van Freeman, director for future workforce development and strategy at the Aerospace Industries Association, moderated the session.

Kim noted that RCCD experienced a 13.8 percent full-time equivalent decline in enrollment, or nearly 5,200 students, for the fall 2020 semester, with declines in enrollment of both Black and white students. While the enrollment of Black and white women declined, however, enrollment of women increased overall. Data from a survey conducted not long after the pandemic began found that 31 percent of the 4,500 RCCD students surveyed were laid off, were furloughed, or had reduced hours, and over half of the students found it difficult to concentrate on school while trying to learn at home. Nearly four in 10 students had no prior experience with online courses, and almost half said they preferred face-to-face learning. At the same time, the effect of COVID-19 on the community has been severe, with over 192,000 jobs lost in the Riverside region by April 2020.

Kim said that according to the *Future of Jobs Report 2020* from the World Economic Forum (WEF, 2020), the time spent on work by humans and machines will be equal by 2025, and in-demand skills for humans will include critical thinking, problem solving, and active learning. That report also projected that 40 percent of all jobs are expected to change as a result of automation, requiring retraining or up-training on the job. "A global pandemic has given us an opportunity to pause and examine what we really need to be doing differently to not only address having machines as coworkers, but also address the systemic and institutionalized systems of racism that have historically plagued our country," said Kim.

RCCD's response has been to use the framework of guided pathways into careers to re-engineer its systems, processes, and practices so that its students are always in the center of education. This means actively developing clear pathways with on and off ramps so that students can seamlessly travel the road to economic mobility via stackable certificates and degrees. One pathway, for example, merges cyber defense and entrepreneurship to cross-train students for business ownership and expand their ability to make business cyber security more effective using artificial intelligence. RCCD has also started programs leading to certificates and degrees in entrepreneurship, data science and analytics, and the gig economy. It has also established an apprenticeship network, a user-focused, navigable apprenticeship system serving the Riverside region's businesses and students.

Kim's recommendations included increasing funding for the National Science Foundation's Advanced Technical Education program that targets community colleges, for Department of Education funds administered under Title III Part F for STEMM programs, and for Department of Education Title V for rebuilding the infrastructure needed to close equity gaps. She also recommended investing in the nation's digital infrastructure and making broadband internet universal, particularly in low-income communities with low rates of college attainment, and implementing workforce incentives that make it worthwhile for both students and employers to invest in retraining and up-training. Finally, she recommended creating stackable and portable degrees similar to the European Union's Bologna Process and its Erasmus-plus program that would allow students to take credits earned to any U.S. institution of higher education.

Increasing the Participation of Women in Entrepreneurship

Torres highlighted how inclusion, social mobility, diversity, and equity are hallmark features at UCR, including in STEMM-related fields, and he predicted that his institution's most innovative contributions to the regional economy are likely to emerge from the interplay of the university's historical strengths in research and education with the pressing need to address challenges posed by population growth and environmental sustainability. All told, UCR has invested $475 million in STEMM-related research, including $3.5 million in a life science incubator complete with wet lab facilities, and various testbed facilities for product and technology validation and for conducting environmental impact research.

Torres noted that half of all start-up companies have no women on their leadership teams and only one in four has a woman among the founders. His institution is addressing this through programs offered as part of its regional small business development center, which is available to all students and faculty at UCR and the region's entrepreneurs. One example is the Women Talk Tech conference that UCR sponsors. UCR's Launchpad program features women founders and social entrepreneurs in its speaker series as well as its Social Innovation Path initiative designed to diversify the pipeline of student entrepreneurs. In response to the dramatic shift to online commerce triggered by the COVID-19 pandemic, UCR now offers courses to help small businesses transition to the online marketplace.

Increasing Diversity in the Healthcare Workforce

As the new associate dean for pre-clerkship medical education, Ford has been working with leadership at the UCR School of Medicine to increase the diversity of California's medical workforce. Out of 299 medical students, 26 percent are Latinx, 7 percent are Black, 31 percent are

Asian, 16 percent are white, 8 percent are of two or more races; 53 percent are women. Since the medical school opened in 2013, nearly 80 percent of the graduates have stayed in California, and 30 percent have remained in the "inland empire." Ford's institution has also worked to increase the diversity of its biomedical science graduate students, and as of September 2020, 23 percent of the program's students were Latinx, 23 percent were Asian, 7 percent were Black, 34 percent were white, and 2 percent were Native American, Hawaiian, or Pacific Islander. Over 60 percent of the students were women.

Ford highlighted the Riverside Bridges to Baccalaureate program, a collaboration between the medical school and Riverside City College designed to provide students of color at Riverside City College with a pathway to research-oriented careers in STEMM. The program's goal is to bolster the ability of students to transfer from a community college into a four-year degree program, and its success so far has prompted Ford to work with Kim to expand the program to all schools in the RCCD. Another program designed to increase the diversity of the biomedical research workforce is a partnership between the University of California system and historically Black colleges and universities.

Despite these successes, said Ford, there are challenges to increasing the diversity of the STEMM workforce. To start, the COVID-19 pandemic is likely to trigger budget cuts that will affect the programs he discussed. On a larger scale, Ford recognized the need for increased local and regional support for these diversity programs, as well as for prioritizing health equity and social justice and addressing systemic bias and racism in institutions of higher education.

Discussion

When asked how the COVID-19 pandemic has affected established workforce training programs beyond potential budget cuts, Ford said the medical program has adjusted well, with students participating in some small classroom settings to fulfill their clinical training. Graduate programs have also adjusted their in-person activities to keep students safe while enabling them to continue their laboratory work. The pipeline program, which provides students with a summer stipend on which they depend for their livelihoods, moved completely online. Torres added that his institution has also adjusted its laboratory-based programs to keep lab occupancy at a safe level, and its programs for entrepreneurs have moved completely online.

His institution's annual small business technology transfer conference was attended virtually by more than 500 individuals.

At RCCD, all 40,000 students attending its three colleges were transitioned to online instruction over 72 hours, which Kim said created difficulties for many of its students who lacked a computing device other than a cell phone, or lacked a broadband connection. External funding, however, allowed RCCD to distribute thousands of computers to its students and set up parking lot wireless internet systems.

Regarding new programs developed in the wake of the COVID-19 pandemic, Kim said that RCCD was well situated as far as programs being offered, and it has embarked on several short-term certificate programs designed to provide new skills to the many students and workforce members in the Riverside area who had been working in service industries. Torres said the same about UCR and noted that the pandemic is likely to have changed the way higher education is conducted in general going forward. Ford said the pandemic has accelerated some of the programs that were already in the works, including a redesign of the curriculum to include interactive learning and smaller class sizes. There has also been an 80 percent increase in the number of faculty who are participating in faculty development programs.

Kim said she would recommend increasing the nation's investment in addressing the digital divide and that the National Science Foundation should increase its funding for advanced technical education that builds middle-skill jobs for high-demand occupations. She would also like to see more equitable investment in higher education, particularly through mechanisms that focus on the STEMM workforce and STEMM infrastructure. Ford seconded the need to make investments in higher education more equitable.

CLOSING REMARKS

Isaac concluded the workshop with a brief summary of the key points he heard throughout the day. The first message that came through clearly was the recognition that expanding and strengthening the STEMM workforce has value for meaningful economic development. It was also clear that the role of collaborative problem solving can maximize resources, reduce duplication of efforts, and increase system efficiencies that will be increasingly important for surviving a near-term funding situation that is likely to be strained. He also noted the emphasis placed by the day's speakers on addressing inequities and the effects of racism and social injustice that so many students experience. For Isaac, this adherence to equity issues will help ensure the success of the programs discussed during the day and increase economic and social mobility among Californians in under-represented groups.

6

Wichita, Kansas

The final workshop of the series focused on Wichita, Kansas, and the surrounding Sedgwick County area, which has grown steadily since the 1960s and become increasingly diverse, with the non-white, non-Hispanic residents accounting for 31 percent of the population. The aviation industry is a major employer in the area, and Boeing's trouble with its 737 MAX aircraft led to layoffs in the region even before the COVID-19 pandemic began. Wichita is unique among the five regions visited in this workshop series in that its economy is tightly connected to one specific sector that has been heavily affected by the pandemic. After reviewing the economic and workforce effects of the COVID-19 pandemic in the Wichita area, the workshop examined how the pandemic shifted Wichita's labor and workforce efforts and strategies, and how the region's higher education and other workforce training efforts responded to the unemployment crisis triggered by the pandemic.

UNDERSTANDING THE ECONOMIC AND WORKFORCE IMPACTS OF COVID-19 ON WICHITA

The workshop was introduced by Donna Ginther, the Roy A. Roberts distinguished professor of economics and director of the Institute for Policy and Social Research at the University of Kansas, and Karen Elzey, associate executive director of Workcred. Ginther said that the poverty rate in the city of Wichita is 14.2 percent, compared to 11.6 percent for the Wichita metropolitan area. She noted that participation in the Supplemental Nutrition Assistance Program has increased by nearly 6 percent since the COVID-19 pandemic began. As was true for the other four cities in the workshop series, average wages in the city are lower than in the metropolitan area, at $32,360 and $34,027, respectively.

Ginther noted that Wichita city is more diverse than the metro area, with a higher percentage of the city's population being Black or Hispanic (Figure 6-1). In the Wichita metro area, close to 31 percent of the population has a bachelor's degree, which makes it one of the more highly educated regions in Kansas (Figure 6-2). Educational attainment in the city is somewhat lower, said Ginther, with fewer residents having at least some college experience.

Unemployment in the metro area spiked to 18.7 percent in April 2020, compared to 11.9 percent for the entire state, with almost 110,000 manufacturing employees filing initial claims as of October 10, 2020. They were joined by 48,000 retail trade employees, 34,000 people employed in other service sectors, 65,000 in hospitality and food services, and close to 50,000 in healthcare (Figure 6-3), five industries that accounted for 65 percent of the people filing initial unemployment claims in the state of Kansas between March and October 2020. At the time of the workshop, the city's unemployment rate had dropped to approximately 5.5 percent, somewhat better than the metro area's 6.4 percent unemployment rate, and the labor force participation rate had fallen by about 2.1 percent since March 2020. Women and Black individuals have fared worse (Figure 6-4). While women

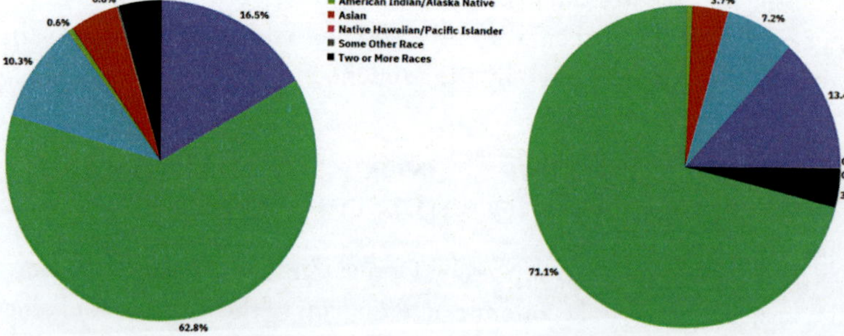

FIGURE 6-1 Demographics of the city of Wichita (left) and the Wichita metropolitan area (right).
SOURCE: Ginther slides 3 and 4.

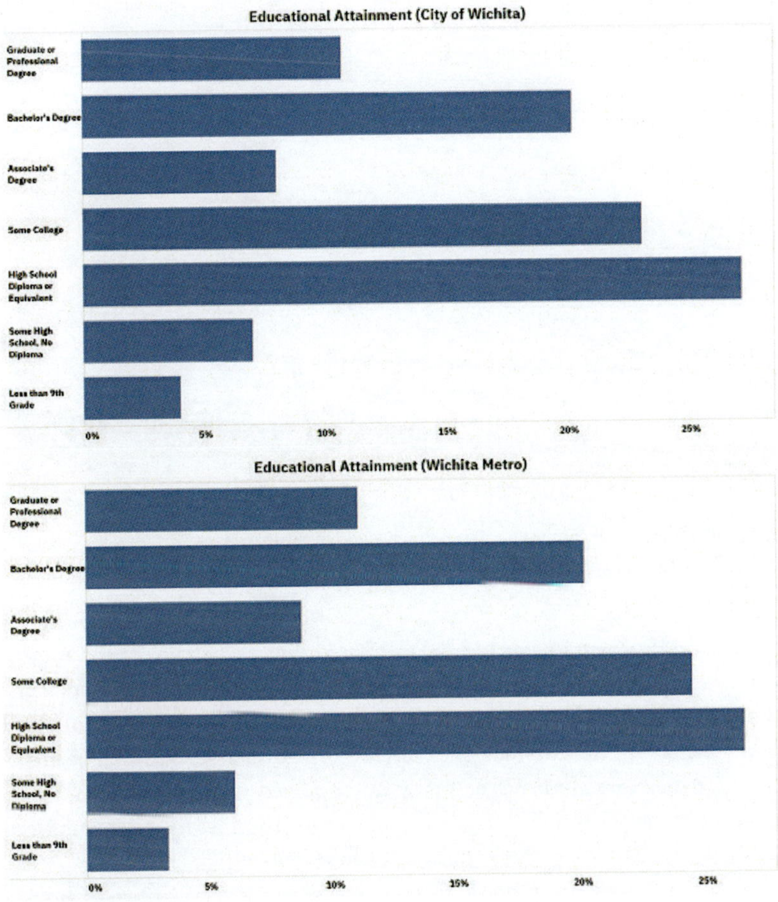

FIGURE 6-2 Educational attainment in the city of Wichita city (top) and the Wichita metropolitan area (bottom).
SOURCE: Ginther slides 5 and 6.

made up 47.5 percent of the workforce, they accounted for 50.7 percent of the claimants. Black individuals made up 5.1 percent of the workforce but accounted for 17.9 percent of the unemployed. Ginther noted that during the week of September 26, 2020, almost four times as many workers in Kansas were on the Pandemic Unemployment Assistance Program as were receiving traditional unemployment benefits.

Ginther noted that the Paycheck Protection Program has been a bright spot for the regional economy. Kansas received some $5 billion in Paycheck

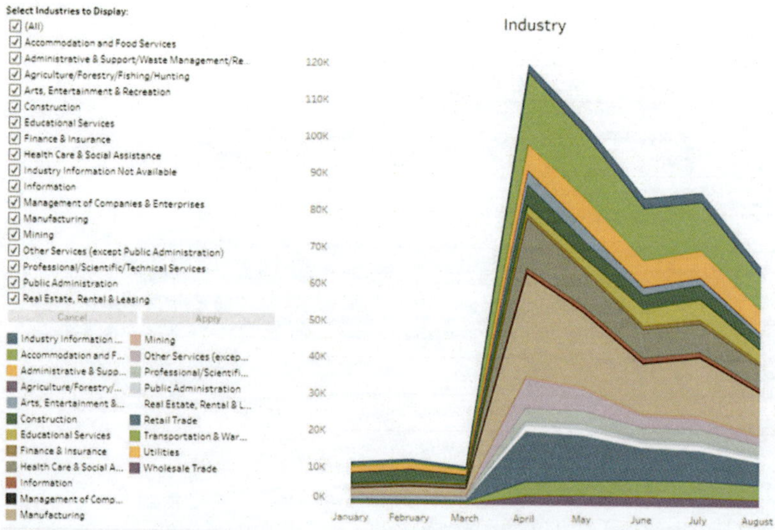

FIGURE 6-3 Unemployment claims for Kansas by industry.
SOURCE: Ginther slide 9.

Protection Program loans, and she estimated that 35 percent of the Kansas workforce benefited from this program. In Sedgwick County, 41 percent of the jobs are tied to this program's loans. In Wichita, 19 percent of small businesses have closed since the start of the pandemic. Ginther pointed out that there are six times as many unemployed workers in the Wichita metropolitan area as there are jobs, and there are 10 times more unemployed workers than there are jobs requiring technical vocational skills.

One unique feature of Wichita is that a greater percentage of households in the city have a broadband internet subscription than in the larger metro area, although a smaller percentage of metro area residents have no internet subscription or access the internet solely through a cellular data plan. A slightly higher percentage of households in the city earning less than $20,000 per year have an internet connection than those in the larger metro area. One big difference can be seen in the percentage of households with no computing device, with 13.1 percent of the city's households in that situation compared to 6.7 percent in the metro area.

According to data from the Atlanta Federal Reserve Bank, approximately 20 percent of the opportunity employment defined as jobs that do not require a bachelor's degree but pay above regional median wages-in the Wichita area is at risk, with the most at-risk opportunity jobs being

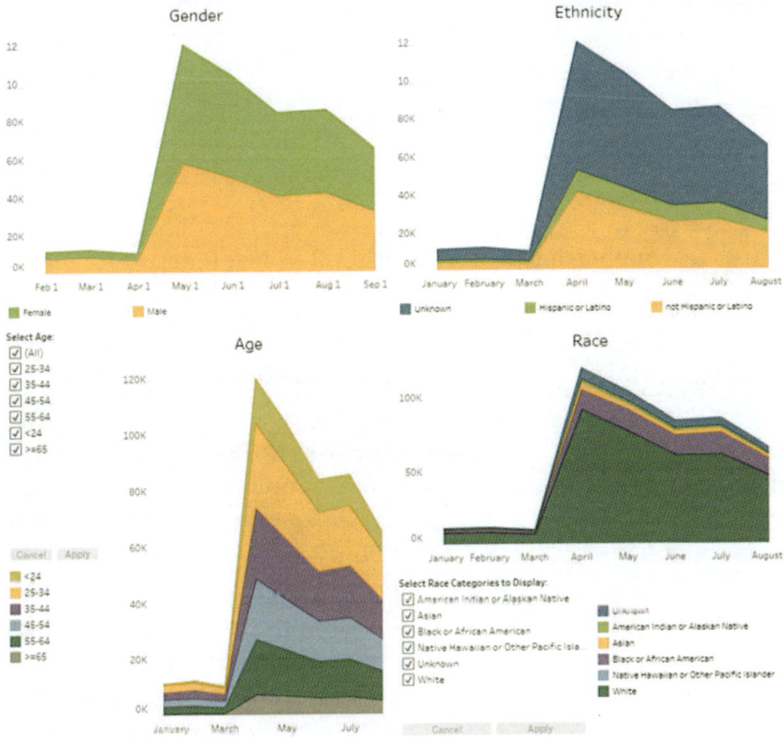

FIGURE 6-4 Unemployment claims for Kansas by age and race.
SOURCE: Ginther slide 8.

secretary/administrative assistant, bookkeeper/accountant, first-line retail clerk supervisor, and carpenter. "There is significant risk to good-paying jobs in the Wichita metro area," said Ginther. She also noted that high-paying jobs in the region (those paying more than $60,000 a year) have mostly recovered, down 2 percent through September 2020, compared to a decrease of 5 percent for middle-wage jobs (those paying between $27,000 and $60,000) and a decrease of 15 percent for low-wage jobs. This situation is similar to what is seen in the rest of the United States.

STATE AND LOCAL POLICY RESPONSES

The workshop's first panel discussed how the COVID-19 pandemic has shifted Wichita's labor and workforce development efforts and strategies. The three panelists were Amanda Duncan, vice president and chief business development officer at the Workforce Alliance of South Central Kansas, a one-stop resource for job seekers and employers; Scott Rigby, assistant city manager and director of development services for the city of Wichita; and Trent Armbrust, director of strategy for bioscience and technology at the Kansas Department of Commerce. Deb Miller, director of the Public Management Center at the University of Kansas, moderated the session.

Supporting the Predominant Regional Industry

The Workforce Alliance of South Central Kansas, explained Duncan, serves as the local workforce development board for the six-county region surrounding Wichita. It operates the Wichita Workforce Center in the city as well as three auxiliary centers in surrounding counties and 10 additional access points. Given the importance of the aviation industry in the region, a wide range of organizations and educational institutions have come together to form the Air Capital Commitment to serve as a coordinated community response to keep and support this industry and its workers. This group has delivered wrap-around services and resources to support laid off and furloughed employees to retain the region's highly skilled workforce, which state officials were worried would be poached by other states. It also lobbied the state to extend state unemployment benefits from 16 to 26 weeks, mainly to enable aviation workers to stay in Wichita, said Duncan. Duncan explained that, thanks to the Wichita Chamber of Commerce, Air Capital Commitment was able to transform itself into the Wichita Resource Center on COVID-19 to extend its assistance to other sectors of the regional economy.

The Workforce Alliance, said Duncan, also secured a $1.9 million Emergency Dislocated Worker grant from the Department of Labor, and the state applied for trade adjustment assistance certifications for business affected by dislocations in the aviation industry. These certifications have enabled Duncan's organization to provide services to 26,000 individuals. The Greater Wichita Partnership, which the organization joined, is now running a retention campaign that promotes retraining as a means of retaining employees in the region.

Rigby's main concern in his position with the city of Wichita is to help local companies upgrade and improve their business units so they can survive the economic fallout from the COVID-19 pandemic and thrive when the pandemic is over. For example, a company involved in aerospace manufacturing pivoted to produce ventilators, which enabled the company to keep over 1,000 of its workers employed and generate an additional revenue stream. The state also helped companies, including some in Wichita, receive grants to manufacture personal protective equipment, which enabled them to expand their manufacturing capabilities and hire more people. He described how in the aerospace sector most companies have been manufacturers, but there may be opportunities for them regarding maintenance, repair, overhaul, and other aspects of aviation.

A Focus on Broadband Access and Increasing Digital Skills

Armbrust described Kansas's new state economic development plan, enacted around the time the COVID-19 pandemic hit. That plan had focused on developing talent for advanced manufacturing as a key pillar, but when the pandemic struck, it became clear that broadband internet access for the small aviation supply chain businesses in the region was a real issue. As a result, ensuring that the state's companies had access to broadband internet and understood how to take advantage of a high-speed connection to improve their businesses became a second key pillar of the state's economic development plan. The state committed $50 million for building broadband infrastructure and another $10 million to help companies and individuals access the internet.

Armbrust commented that Kansas has also allocated funds to enable its technical and community colleges to purchase the equipment they need to train their students and upskill employees on the latest digital technologies, including cyber security. The state has established a program that will quickly train hundreds of people to obtain a certificate in information technology, cyber security, and information technology project management, all jobs in high demand in Kansas. The state has also been successful leveraging its investment in broadband access and training for information technology–related jobs to bring some aerospace supply chain companies that had moved overseas back to the region. In addition, he said the state has included a 10 percent boost on its working capital grants for those business that are owned by women and people of color.

Discussion

Regarding programs started during the COVID-19 pandemic that might end up continuing once the pandemic ends, Rigby said that one program likely to survive enables restaurants to apply online for permission to expand their operations outdoors and receive an answer within 48 hours, rather than the weeks or months it took prior to the pandemic. He noted that the city has also looked at accelerating other application processes and reducing red tape now that it has seen that it can in fact do so when pressed. Working with the county, there is now a small business grant of $5,000 available to help companies pivot quickly when the situation demands they do so to survive.

Duncan said her organization has integrated more thoroughly with the Kansas Department of Commerce, something that will continue. It also worked with the Greater Wichita Partnership to establish a program for advertising the availability of public sector jobs, something that public agencies themselves are prohibited from doing. The Workforce Alliance will also continue its new efforts in retraining employees for information technology, automation, and healthcare jobs that includes an aggressive outreach effort to recruit people into those jobs.

One thing the state has learned through the partnerships it has built during the COVID-19 pandemic, said Armbrust, is the urgency of repositioning the state's workforce to take advantage of new jobs in automation and information technologies. As a result, the state is going to be "laser focused" on developing digital skills in its workforce so that it can remain competitive on a national and global scale. Rigby noted the newfound awareness among leaders in Wichita and the surrounding area that it should accelerate efforts to diversify the economy so that it does not rely so heavily on aerospace manufacturing or even healthcare. He also pointed to the city changing its incentive guidelines to focus more on small and mid-sized businesses and enable them to grow.

When asked whether producing personal protective equipment will become an ongoing industry in Kansas or if that will be a short-term occurrence, Armbrust replied that he believes some of the companies making this equipment will continue to do so going forward, and thanks to money from the Coronavirus Aid, Relief, and Economic Security (CARES) Act, they have been able to invest in technology that makes them competitive with foreign manufacturers. One company, for example, used CARES Act funds to make robots that help make personal protective equipment, and that firm will be able to use the technology it developed to make equipment for the pharmaceutical industry and others.

Rigby noted that the speed with which the pandemic has forced the city to work will benefit the city and its residents for years to come. His hope, though, is that the changes in the work environment that the pandemic has triggered will lead to companies thinking differently about where to locate their operations; for example, the cost of living is much lower in Wichita than New York or Boston. Duncan seconded that idea and added that she wants local employers, when the region emerges from the pandemic, to feel they were supported and encouraged to grow and create job opportunities for young Kansans who might then stay in the state instead of migrating to the coasts for jobs. Armbrust sees the state's renewable energy portfolio expanding and foresees being able to recruit talent because of the sustainability efforts that companies are making. And he hopes that the state will not wait 24 years between efforts to develop an economic development plan, which was the length of time since the state first developed such a plan.

RESPONSE OF WICHITA'S HIGHER EDUCATION WORKFORCE TRAINING EFFORTS TO THE COVID-19 CRISIS

The workshop's second session highlighted changes in higher education workforce development programs in the Wichita region that were triggered by the COVID-19 pandemic. The four panelists were Susan Norton, director of adult learning at Wichita State University; Kimberly Moore, director for workforce, professional, and community education at Wichita State University; Christopher Stanyer, vice president of career services at Goodwill Industries of Kansas; and Deltha Colvin, associate vice president for special programs in the Office of Regional Engagement and Economic Development at Wichita State University. Karen Elzey, associate executive director of Workcred, moderated the session.

Wichita State, explained Norton, had 10 days to move from a normal classroom setting to totally online, and fortunately the university had enough of a technology infrastructure to make it happen. That did not mean the transition was easy, given that there was a significant learning curve to deliver course material via Zoom. Staff from the university's media resource center and instructional design team were incredibly helpful in that regard.

Moore, who oversees the university's Badges program, said there was no transition for her project because its educational programs focused on workforce skills and competencies were already exclusively online. The Badges program, launched in 2015, had 954 individuals enrolled in 2019—mostly non-degree-bound working professionals—and offered 83 different badges that are defined as an academic course worth one undergraduate or graduate credit or less. The self-contained, ready-to-go courses that individuals can take to upskill or re-skill from the comfort of their own homes are graded on a pass-fail basis, and students can sign up for a course at any time.

Before the COVID-19 pandemic struck, the "badge" topics were faculty-driven, but since the pandemic began, Moore and her colleagues have focused on building business and industry partners that will drive curriculum development and instructions. One partnership with direct support professionals, for example, provides home-based technology services for older individuals. Another partnership, said Moore, is with regional nonprofit organizations that employ and provide services and educational opportunities for people with intellectual and developmental disabilities. A similar partnership with the National Association for the Dually Diagnosed focuses on individuals who already have degrees and are upskilling to work with those who have intellectual and developmental disabilities.

Moore reported that the team is also working with Working Men of Christ Ministries to provide and develop badges that individuals recently released from prison can use to get the competencies they need to find a well-paying job in the community. Another program, developed with the help of the National Association of Credit Managers, aims to provide additional training for finance graduates as a means of getting more young people into that profession before the large tranche of 50-plus-year-old credit managers retires over the next decade or so.

Stanyer described some of Goodwill's programs designed to help individuals overcome barriers to employment. The NextStep Alliance, Goodwill's adult education program, is a partnership with Wichita State. Its Learn IT Lab, funded by a three-year grant from Google, can provide digital skills to all Kansans throughout the state using a recreational vehicle as a mobile classroom complete with computers and internet access. The third program, NextStep Unlocked, provides high school credit for completing technical college courses and is intended to help people released from the criminal justice system rebuild their lives.

When the pandemic hit, Goodwill made use of its online programs, but Stanyer and his colleagues are having to work hard to have a one-on-one connection with the students, most of whom come from a high-need demographic with few resources. Strayer reported that by using CARES Act funds, Goodwill has been able to set up a computer laptop loan program. However, the Learn IT Lab was grounded because nobody wanted to be inside an enclosed classroom, and it is only now relaunching as an onsite training program conducted at an employer's location at no cost to the employer.

Colvin explained that all of the programs supported by Wichita State University's Office of Regional Engagement and Economic Development are funded through nationally competitive grants. Several of these programs encountered significant difficulties once the pandemic hit, particularly those that required students to learn communication programs such as Zoom and Microsoft Teams and to have a computer. In addition, some programs needed to provide new forms of support for students, such as providing meals.

Discussion

To help individuals who did not have reliable internet access, Goodwill was able to buy 250 computers and set up a computer loaning program. This, however, did not solve the access problem, because the price of connecting program participants on cellular data plans would have been astronomical. Stanyer is hoping that the state will pass upcoming legislation that would provide funds for extending broadband access into lower socio-economic status neighborhoods. Colvin said the Department of Education has allowed her program to buy computers and internet hubs to provide access to students that would otherwise be unable to participate. The bottom line, she said, is that if the students in her program do not have internet access, they cannot learn.

Norton, responding to a question about the types of supports available to help instructors better understand the technologies they use to teach online, said that Wichita State University was fortunate to have an instructional design group that can help faculty with their online issues.

In terms of working with the Badge program to align with the needs of employers as the economy changes in Wichita, Moore said that she and her colleagues are engaging with human resource managers, hiring managers, and executives and directors at a variety of local employers to determine what jobs are open and what skills are lacking in the local workforce. One thing she has been hearing is that job applicants, and even some of her program's graduates, are lacking soft skills, so her group is now developing badges in the area of time management, communication skills, teamwork, and the ability to work in a digital environment. Moore has also been working with national associations that have their finger on the pulse of future job projections so that her program can stay abreast of the needs of a changing economy.

Norton said she has been hearing similar things about adult education, particularly the need for digital skills and digital literacy. She noted that there is a strong partnership among the programs run by everyone on the panel: "We are trying to use these relationships and programs to make it easier for people to increase their skill sets and increase their job opportunities and find a path to where they want to be."

Responding to a question about how the partnership between Google and Goodwill came about, Stanyer said that Google approached Goodwill International in 2017 looking for ways to invest in digital literacy across the country as a means of preparing people for the modern workforce. He noted that when the three-year grant from Google ends in December 2020, Goodwill will renew it because it has become an important component of workforce training, and it will be needed to supply workers with digital skills for Amazon's new facility in Wichita.

At the conclusion of the session the panelists were asked to list one action needed to assist them in their roles in helping meet the demand for the science, technology, engineering, mathematics, and medicine workforce. For Norton, a significant challenge is keeping up with how fast the job market is changing and what employers need from their students regarding skills, particularly during the pandemic when face-to-face meetings are not happening. Colvin said she would like to have a cadre of volunteers who would be willing to serve as role models for her students and provide one-on-one help becoming computer literate and learning new digital skills. Moore said that communication and partnerships with local employers is key to identifying the specific skills and competencies they desire in their current and future employees. Stanyer also wanted synergies through partnerships to help boost adult education in Kansas, particularly given that budgets will be tight for some time because of the economic effects of the COVID-19 pandemic.

RESPONSE OF WICHITA'S WORKFORCE TRAINING INFRASTRUCTURE TO THE COVID-19 CRISIS

The workshop's final panel discussed post-COVID-19 policy responses and investments in workforce training programs, including the effect of the pandemic on workforce alignment and learner access. The panelists were Michele Gifford, human resources director for operations and workforce programs at Textron Aviation; Amy Williams, senior manager for global community relations and corporate administration at Spirit AeroSystems; and Cheryl Childers, human resource manager at Cox Machine. Lida Beninson, senior program officer at the National Academies, moderated the session.

Gifford explained that Textron Aviation is the largest general aviation and business jet manufacturer in the world, with skill sets in its employee base including accountants to welders and everything in between, including highly educated engineers to true craftspeople on the shop floor. Williams said that Spirit AeroSystems is a tier one manufacturer and supplier to Boeing and Airbus, as well as to defense manufacturers. As with Textron Aviation, Spirit AeroSystems has a workforce spanning highly trained craftspeople on the shop floor to highly educated engineers, plus the sales and administrative staff. Childers explained that Cox Machine, a family-owned company, is a supplier to the Textrons and Spirits of the world, as well as companies involved in the space industry, and, like them, it employs a wide range of skilled and educated workers.

Discussion

Williams commented that for the five years she has been at Spirit, the company was going "like gangbusters," as was the rest of the aviation sector in Wichita. The result was the development of partnerships throughout the region to grow talent and encourage professionals to move to Wichita. Then the local aviation industry took back-to-back hits, with Boeing's 737 MAX troubles and the COVID-19 pandemic. The challenge became one of retaining the workforce, so the company let its employees do volunteer work in the community to make sure they were getting to do something meaningful. But that could only go on for so long, prompting the company to start thinking about diversification.

Then, out of the blue, Spirit AeroSystems was given the opportunity to make ventilators, and the company deployed close to 700 employees to that new project. The point, she said, is that organizations, no matter how long they have existed and no matter what they have been doing, should be open to new opportunities and be ready to seize them based on the internal skills, strengths, and competencies.

Childers said Cox Machine has been concentrating on training for the past three to five years to fill the huge skills gap in Wichita. Her company developed partnerships with Duncan and her team at the Workforce Alliance, as well as with Wichita State and the university's applied sciences and technology community college campus (Wichita State University Tech), to build the local workforce so that the company would not have to leave the Wichita region. In fact, she said, many of the local manufacturers, including Textron and Spirit, worked together on the challenge of developing the local workforce.

She noted that, as a smaller company, its training efforts needed to be broader in scope, and over the past six to nine months, the company's training efforts have increased so that once the economy turns around, it will have the workforce ready to take advantage of new opportunities. Some of the funds to support this effort came from the Paycheck Protection Program, and some came from the company's reallocation of funds to emphasize training.

Gifford said that three years ago, the company began investing heavily in the region's high school students to get ahead of the retirement bubble that will hit in the next five to seven years. Her company is also working on developing apprenticeship programs with Wichita State University Tech, and it has a summer externship program that brings college and career counselors and career and technical education teachers into its aviation factories so they can sell their students on a career in the aerospace industry. She noted, though, that the pandemic forced Textron to put its summer internships for high school students and its externship programs on hold, since it was not feasible or practical to bring people into the factories. Assuming some sense of normalcy has returned by summer 2021, the company plans to restart those programs. In the meantime, it has established virtual shadowing opportunities for local high school students as an approach to keeping students excited about a possible career in the aerospace industry.

Williams commented on the critical role of partnerships throughout the community in supporting the mass of employees who have been affected by the Boeing MAX 737 troubles and then the COVID-19 pandemic. Childers also commended all of the community partners that have come together to help the region's employees who have been hit hard by the devastating effects of the pandemic on the airline industry, and hence, on aerospace manufacturers. She also noted how important civic pride has been throughout the pandemic. This attitude has manifested itself in the way that companies that normally compete have banded together to help keep the region's workforce intact for when the aviation industry rebounds. The only way that happens, she continued, is to have strong relationships built throughout the community, and she implored those attending the workshop who might live elsewhere to start building those relationships now. "Contact United Way, contact your workforce alliances, contact your local schools and trade schools," Childers said. "Get those relationships going, because when things go south, you are going to have to rely on those people, and you are going to have to help each other out."

When asked what recommendations she would make to her company's partners or to local, state, and federal officials that would help support its workers, Williams first noted that the Wichita Promise program pays for living expenses while workers are taking classes and has been successful at helping workers get the retraining they need to either move into a new job or be prepared to move up in their old job once the aerospace economy starts recovering. She would like to see more support for this type of program and a more concerted effort to promote this type of opportunity, not just in Wichita or Kansas, but across the nation.

Childers said she wants to see funding restored so that schools can once again offer shop and automotive classes and start students down a path toward the skilled trades that offer well-paying jobs. She noted that because schools have had to cut these programs, local firms are having a hard time finding people right out of high school to fill middle-skill jobs without first having them go to community college. Gifford seconded Williams's and Childers's recommendations and added that she would like to see reduced barriers to offering work-based learning credits for high school students. She also recommended steps that might make it easier for smaller businesses to offer apprenticeships.

CLOSING REMARKS

To conclude the workshop, Beninson summarized some of the common themes discussed throughout the day. Wichita, she began, has been hit particularly hard by the pandemic and the Boeing MAX 737 grounding, which has been devastating for a large swath of Wichita's workforce, and the companies in the region have taken this "opportunity" to think about how they could expand, diversify, or pivot. This, she noted, would not have happened when the local economy was booming and companies were working at full capacity to meet customer demand. However, expanding, diversifying, or pivoting to new product lines requires new skills among the local workforce, and local educational and workforce training programs have responded with numerous programs to provide that training, often in partnership with local companies.

Access to reliable broadband internet is a critical resource, and the Wichita area has many gaps in broadband access that it is trying to address. At the same time, the community has recognized that there are some people who need not only internet access but also training in how to operate in the digital environment, and Wichita is developing programs to address that need. In addition to addressing the digital skills gap, local programs are also coming online to provide training on "soft skills" such as teamwork, time management, and communication that local companies have said are lacking in some new employees. Beninson noted how hard Wichita is working to keep its skilled workers and keep them engaged in the community. Public-private partnerships have been key to this effort, as has local civic pride around Wichita and the quality of life it can offer.

7

Regional Observations with National Implications

The final workshop in the six-part series aimed to reflect on the adaptability, challenges, and needs of workforce training efforts in science, technology, engineering, mathematics, and medicine (STEMM) across the nation in the wake of the COVID-19 pandemic and how regional successes, discussed in the previous five workshops, can inform efforts across the nation. As Jeff Bingaman, former U.S. Senator from New Mexico, noted in his introduction to the sixth workshop, the nation is living in a tumultuous time that has disrupted countless aspects of everyone's lives. For more than 11 million Americans, as of October 2020 that disruption includes job loss, which is both a tragedy in terms of lost wages and family stress and an opportunity to expand STEMM workforce training to meet the demands of 21st century employers.

This series of workshops, said Bingaman, aimed to understand how the COVID-19 pandemic has affected the nation's skilled STEMM labor needs and how different regions and their workforce training programs are trying to meet these needs. The committee that planned these workshops sought to understand the different ways in which cities have responded to the labor shifts brought on by the pandemic and how cities are adapting their workforce training efforts to help displaced workers move into skilled jobs in which they can earn a living wage capable of supporting a family. After reviewing the discussions from the first five regional workshops, the committee recognized key themes across the workshop series to both highlight and discuss further in a sixth and final workshop. These themes include the digital infrastructure needed to equitably support workforce training, the key services needed to support equitable access to workforce training, and the future design of workforce training.

REGIONAL STEMM WORKFORCE TRAINING RESPONSES IN THE WAKE OF COVID-19

In this workshop's first session, three members of the workshop planning committee shared their views of common themes and key differences across the five cities highlighted in the preceding five workshops. The three committee members were Maria Flynn, Donna Ginther, and Rachel Lipson. Karen Elzey, another committee member, moderated this panel.

Lipson noted the complexity of the challenges the nation is facing. "At the precise moment at which we have a huge influx of folks into the unemployment system and of folks who are looking for new jobs and potentially in search of new training opportunities, we are putting enormous stresses on our system," said Lipson, who then asked, "How do we find, contact, reach those folks, help them navigate through a complex system, and help them find opportunities that make the most sense for them at this moment?" Her main takeaway from the preceding five workshops was the multi-pronged nature of the efforts undertaken by all of the highlighted cities—for example, efforts in Boston to put paper flyers in public housing and convenience stories—to engage with unemployed workers, particularly around outreach and enrollment into new programs. Lipson added that one of the big challenges now is balancing the more traditional in-person services, such as career navigation and coaching, with access to the digital skills people need in the changing U.S. economy to secure well-paying jobs. She credited the five cities with trying innovative new approaches to addressing this challenge.

Flynn found the stories of the five communities compelling and energizing because of the innovation that is happening in those five regions as they adapt to the unique needs of their regional labor markets. She reiterated the challenge of connecting workers in industries devastated by COVID-19, such as the tourism, restaurant, retail, and hospitality industries, with STEMM workforce training programs that can provide opportunities to these workers in industries with a growing demand for a skilled STEMM workforce, including technology, healthcare, and certain manufacturing sectors.

Ginther was interested in how these five communities all had strong grassroots efforts to meet the needs of people affected by the COVID-19 pandemic. She found it encouraging that so many community-based public-private partnerships sprung up immediately after the pandemic hit and during the subsequent recession it has caused. At the same time, she was struck by the data showing the disproportionate impact of the current economic situation on women and people of color who had low-skilled jobs in the service sector. The challenge for community organizations is to help those workers move into sectors with high demand and higher wages, such as healthcare and digital technologies.

Elzey asked the panelists to talk about what has changed during the pandemic regarding who is delivering workforce training. Flynn cited how community colleges have stepped up in some interesting ways, and said she was also impressed with how rapidly many workforce training organizations flipped to online delivery and that virtual online offerings will continue once the pandemic is over. She also noted that some community-based organizations are starting to partner with one another to expand what they can do for the local unemployed or under-trained workforce.

When asked to comment on the challenges of being able to deliver services virtually, Ginther highlighted one common theme throughout the workshops—the need to greatly expand broadband internet access and make it affordable for low-income families, as well as establish mechanisms such as lending libraries that will enable trainees to work on a laptop or tablet. "You cannot take workforce training on a mobile phone," said Ginther. She pointed out that people who could afford broadband access and who could work at home have been able to get through this crisis relatively unscathed. In contrast, those who could access the internet only on a mobile phone have not been able to get the training they need to transition from a job that has effectively disappeared to a job of the post-COVID-19 future.

Lipson commented on the professional development activities that were required to bring faculty and staff up to speed in terms of their ability to teach and train people in a digital environment. She considered this an area that merits more attention going forward. In the same vein, she noted the need to help trainees adjust to a virtual learning environment and to develop mechanisms to allow feedback from trainees to ensure that online training is working for them.

Partnerships to Get People Back to Work

Next, Elzey asked the panelists to list some of the highlights they heard regarding the ability of partnerships to get people back to work.

Ginther thought that Pack Health, in Birmingham, was a good example of a company seeing that it needed trained workers to help manage its clients' chronic health conditions and responding by developing its own training and certification programs. "This was a dynamic and exciting model that may be useful for other industries and sectors, as we move from the service-sector jobs that are going away into higher-skilled healthcare and technical jobs," said Ginther, who noted that those individuals who went through the online training landed well-paying jobs.

Flynn said she heard that employers were eager to move away from the four-year degree as the traditional hiring signal and toward a system of stackable, industry-recognized credentials. One perhaps unappreciated aspect of credentials and badges is that they allow people who have lost their jobs and cannot afford to stay out of the workforce for very long to get the training they need quickly and rejoin the workforce. The challenge, she said, is finding more models that integrate learning and work so that people can be sustaining themselves and their families while they are earning that next credential. Flynn was also impressed by institutions of higher education creating wireless internet hotspots in their parking lots to create broadband access for students who may not be able to afford it at home.

One program that caught Lipson's attention was the Virginia Ready Initiative, which was able to leverage an existing infrastructure of statewide initiatives and work closely with the Virginia community college system. She also highlighted the Virginia Fast Forward workforce credential program's efforts to help people navigate the workforce training system to find their way onto pathways that will lead them to credentials and new job opportunities while also providing them with a financial incentive to complete the program.

Matching People to Jobs

Elzey asked the panelists to address the challenge of matching people who are trying to identify the real job opportunities that exist and the trainers who are looking for individuals who will fit well in their programs. She also asked them about the types of data used by communities to better understand what a community's needs are and create the appropriate match that can help both individuals and businesses be successful. Flynn noted that Birmingham used data from Burning Glass, a provider of real-time labor market information, to help guide its workforce training efforts. She also noted the general lack of good career navigation help in the United

States, both for workers and students. This is where community-based organizations can and do help, but the key is moving those types of programs to a larger scale.

Ginther said that detailed weekly unemployment claims can provide useful information given that those data are broken down by education, gender, race, and industry. She also pointed out that while workforce training efforts are leaning heavily on community colleges, enrollment is down at those institutions. "The people who need re-educating the most do not necessarily have the resources to invest in their education," said Ginther. Her suggestion was to provide more earn-to-learn opportunities so that people can support their families while they are retraining.

One story Lipson found interesting was about the new hospitality center that North Essex Community College built after years of planning and fundraising only to have the food industry crash during the pandemic. For her, that raised an issue that she felt received less attention than it might have, namely, how to give people transferable skills that enable them to move across different types of occupations and roles in the workforce. In her opinion, the challenge is to provide people with skills that will be durable and resilient and that provide them with the strongest foundation possible so they can shift between jobs and even industries as the job market changes.

Imparting new skills, said Elzey, is just one part of the equation to help people move up in the world, with the other part being the social services that many people are in need of, such as childcare and transportation. Flynn noted that the pandemic has brought the conversation about supportive services into the mainstream, which was long overdue, and her hope was that it would result in more money being spent to meet the transportation and childcare needs of an increasing number of people. At the same time, there is a need to help people navigate the support services sector, which often involves going from agency to agency to access these services. "How can we be helping states and communities to streamline these efforts and make sure that resources are getting to where they are most needed?" asked Flynn.

Rethinking the Nation's Social Safety Net

Ginther commented that, given the extent of unemployment and the length of time that people are remaining unemployed, policy makers need to be rethinking the nation's social safety net. "COVID is a natural disaster, and we depend on our government to provide social insurance in the face of a natural disaster," said Ginther. The problem is that state governments are largely responsible for providing these services, but their budgets are under pressure because of the pandemic. While the nonprofit sector has stepped up, the need far exceeds its capacity to meet all the needs that exist today. "We need to think about a larger national strategy to provide support to people who lost their jobs for no reason, no fault of their own, as we navigate the COVID-19 pandemic," she said.

Lipson wholeheartedly echoed Ginther's comments and added that there is a need to rethink what the standard set of employment benefits should be going forward. For example, with more people working at home, the traditional benefit packages that include allowances for transportation and parking might not be as relevant as they were pre-pandemic. Perhaps, she said, those funds could go toward expanding childcare, mental health services, or employee assistance programs. Her hope is that companies will lean into providing more help with the latter set of services.

Elzey then asked the panel whether there is a need for a national effort to illustrate the importance of developing a STEMM workforce and whether the COVID-19 pandemic has changed how people view a STEMM workforce. Ginther stated that a STEMM workforce is essential for the future of the nation. For example, automation will replace jobs, but there will be new opportunities to create the technical skills to program the robots and fix them. In her opinion, the displacement triggered by the pandemic has brought the future closer, increasing the urgency of shifting jobs away from the service sector to those that require more technical skills. Flynn said she would like to see the importance of STEMM jobs pushed earlier into high school or even middle school as well as among career counselors who work with the unemployed. An equally important piece is enhancing access to high-quality, work-based learning opportunities. Lipson added that it will be important when thinking about getting people into growth industries, such as healthcare, to help individuals determine the types of jobs in which they will thrive. For example, training someone for a career in healthcare may not fit that person's interests, strengths, or feelings of safety.

Potential Policy Levers at the State and Federal Level

Turning to the subject of state and federal policy, Elzey asked the panelists to talk about the policy levers that could be used to address some of the challenges that emerged throughout the five workshops. Flynn replied that policy makers need to come up with new approaches to financing workforce training, including broadening Pell grant eligibility to include short-term credentialing or providing more flexibility to the Workforce Innovation and Opportunity Act in terms of eligibility to participate and how it allocates training dollars. And even with those changes, she said, new alternatives are needed to best support students and dislocated workers. Lipson seconded those suggestions and added that the federal government should provide funding that will enable all 50 states to rebuild and enhance their existing digital infrastructure to reach everyone in this increasingly virtual and remote world. Lipson expressed hope that once the pandemic is over, the nation will take the opportunity to step back and think about the lessons learned during the pandemic as they concern workforce training and the provision of wrap-around services.

In general, said Ginther, there is a need to reduce administrative barriers to getting wrap-around services. In some states, applications for services can only be done online, while many people who need those services have only a mobile phone. She noted that 10 percent of the K-12 students in Kansas do not have internet access at home. Seeing internet access as on par with other utilities and not a luxury in today's world, Ginther said the federal government has an opportunity to increase the nation's broadband infrastructure, and in doing so, bring down the price so that every student and every displaced worker can afford internet access. "We need a comprehensive broadband policy for the whole country," she added.

To conclude this panel, Elzey asked each panelist for one takeaway message. Lipson said she would highlight some of the innovations that are happening at a state and local level in this crisis environment. "It was so exciting to see the ways in which these regions and industries were iterating, adapting, and figuring things out in the moment and being flexible to try to help as many people as possible," said Lipson. The next stage, she said, will be to figure out how to scale those innovations so that more people can benefit. Ginther said she could imagine the federal government providing a pool of grant money, to be administered by the states, that allows for providing enhanced social services and wrap-around services to support those programs that are taking root right now. "What struck me was how many local targeted innovations were occurring in all of the regions that we looked at," said Ginther. "We just need to support them, get out of their way, and help them to support the people who are most in need."

One thing that struck Flynn was the change in mindset among local leaders from "I can't" to "I must." She said it is important to remember that the work that is really happening is at the local and regional level, making it important to find those local success stories, help policy makers understand what is playing out in those communities, and help communities learn from one another.

DIGITAL INFRASTRUCTURE TO EQUITABLY SUPPORT WORKFORCE TRAINING

In the workshop's next session, three panelists discussed efforts to improve nationwide broadband access so that displaced workers can participate in virtual workforce training programs. The panelists were Matt Dunne, founder and executive director of the Center on Rural Innovation; Laura Spining, director for broadband infrastructure at the Department of Commerce's National Telecommunications and Information Administration; and Nicol Turner Lee, senior fellow in governance studies and director of the Center for Technology Innovation at the Brookings Institution. Rick Seltzer, projects editor at Inside Higher Ed, moderated the session.

The Center of Rural Innovation, explained Dunne, is an action tank dedicated to ensuring that rural communities can succeed in the digital era. The rural opportunity gap emerged in 2008, driven by automation and globalization. Even as of January 2020, over half of the nation's rural counties had not recovered to their pre-2008 recession levels, and all of the difference between the urban and rural economic recoveries can be accounted for by the digital economy coming to urban and not rural areas. Dunne said that job loss arising from automation hit rural areas disproportionately.

Today, during the COVID-19 pandemic, the rural digital divide is affecting the ability of children to participate in learning, and people who might have the opportunity to work at home cannot because of limited broadband access. Addressing the rural digital divide, said Dunne, is a critical piece of ensuring that rural areas recover from the COVID-19 pandemic in better shape than they did after the 2008 recession. Spreading broadband access to every area of the country will be as important to rural economies as was the nation's effort to electrify the entire nation in the early to mid-20th century.

Lee, who is writing a book on the U.S. digital divide, noted how digital access corresponds and correlates with the level of opportunities available to people. The disparate effects occurring in areas of the country as a result of a lack of internet access have much to do with systemic inequalities that were present before the COVID-19 pandemic. Lee explained that when the pandemic began, some 80 million people had no internet access, and this number has likely risen given the economic devastation that has resulted in unemployment, loss of income, and even evictions. Even worse, she added, of the 50 million children sent home from the nation's schools, between 15 and 16 million do not have broadband access or a computing device, and out of that number, 9 million are Black and brown children. These are the nation's future workers, and if future jobs depend on being digitally literate, these children will be disadvantaged going forward. "We have to pay attention to this issue, not as something that is marginal, but that is particularly important in ensuring that the other America, the America that probably does most of the work in this country, is included in getting the benefits of digital opportunities," said Lee.

Spining, whose agency has been working on ways of addressing the digital divide, said she has seen a dramatic increase since the COVID-19 pandemic began in an appreciation of just how urgent it is for the future of the nation to make broadband access universal. Her agency is leading the coordination of the different aspects of broadband funding that the federal government is undertaking across multiple agencies. Most of that funding comes through the Federal Communications Commission, and the U.S. Department of Agriculture's rural utility service program funds broadband deployment. The Department of Housing and Urban Development has recently increased its funding of broadband programs as well as training sessions to help students and unemployed workers gain digital skills. States, too, have recently boosted their broadband and digital inclusion programs.

KEY SERVICES THAT SUPPORT EQUITABLE ACCESS TO WORKFORCE TRAINING

The next panel discussed higher education and social service strategies to connect displaced workers with training opportunities to meet local STEMM employment needs. The three panelists were Tamar Jacoby, president and chief executive officer of Opportunity America; Rebecca Rowley, president of Santa Fe Community College; and Paige Shevlin, director of policy and national initiatives at the Markle Foundation. Scott Gullick, senior director of Nest, YearUp's research and design laboratory, moderated the session.

The challenge that Jacoby sees is not just the number of Americans that have been thrown out of work by the pandemic, and not just the inequitable skew of the people thrown out of work, but also an economy that is changing so fast that it is hard to keep up or make good decisions about how to help people get or prepare for jobs. "Which jobs will be eliminated, and which will be growing?" she asked, "and most importantly, how is what we used to call the 'future of work' going to transform the economy in years ahead?" The COVID-19 pandemic's economic shock has spurred a quantum leap in digitization, automation, robotics, and artificial intelligence, aspects of the future of work that was thought to be a decade away but happened shortly after the pandemic began. "No one knows where the economy is going, and that makes it hard to prepare people for jobs," said Jacoby. "There is certainly no point in training people for jobs that are going to disappear."

When thinking about the millions of Americans who are going to need job-focused reskilling and upskilling in years ahead, it will be important to identify the institutions that can provide this, said Jacoby. To her, no institution is in better position than the community college to provide that training given their infrastructure, instructors, training facilities, and, most importantly, reach and scale. She noted that few Americans live more than a 30-minute drive from a community college, and that community colleges were educating nearly 12 million people a year even before the pandemic struck, compared to 18,000 in boot camps and 225,000 in government job training programs. While fulfilling their promising role in training Americans for the jobs of the future will require at least some change at community colleges, the good news, said Jacoby, is that the necessary changes are already underway at many of these institutions and have been for over a decade.

Going forward, Jacoby sees the need for several ingredients to prepare the nation's workforce for the future. The first is sophisticated, up-to-date labor market information. The current survey-based government data are not of much use anymore because the world is changing too quickly. Second on her list is robust employer input. "You cannot do job-focused education and training without input from people who know the jobs, who know how their industries are changing and what skills are in demand," she explained. While most community colleges have relationships with employers, too many of them emphasize quantity over quality, which translates into less informative and less useful collaborations.

Third, dislocated workers need better strategic navigational supports to help them identify skills they have that can be useful in a different industry. In Jacoby's experience, too many unemployed workers make bad choices about training and often have difficulty sticking with it. The fourth, and perhaps most important, item on her list is changing the current policies and laws that often require learners in short, job-focused programs at community colleges to pay for those programs themselves. In general, she said, individuals cannot use Pell grants to pay for those types of programs. Moreover, the training dollars available through the public workforce system do not come close to meeting the demand for training. Community colleges cannot be expected to provide training for free. The final item on her list is job placement metrics. "I believe that public funding for job-focused upskilling should be contingent on success in getting people jobs," she said, a position she acknowledged is controversial.

Rowley began her comments by noting that community colleges see themselves as the economic generators in their communities, whether they are rural communities or major metropolitan areas. She noted that the COVID-19 pandemic has created several opportunities for community colleges, including the capacity to build stronger online programs. The pandemic has also forced these institutions to collect better data on what students need now.

In New Mexico, there is severe need for more healthcare professionals, and Santa Fe Community College is focusing on trying to ensure that its healthcare students can complete their programs on time as much as possible. To do that, Rowley's institution is offering more clinical classes in simulation format than it would have ever done before the pandemic. She noted that this has been made possible by policy changes at the state level, as well as her school's prioritization of small group, face-to-face instruction in its healthcare curricula as well as several other STEMM areas, to meet immediate workforce needs. She pointed out that her institution has experienced an across-the-board enrollment decline during the pandemic, but that there has been a much larger decline in its Native American student population, likely because of a lack of internet connectivity.

As many speakers throughout the workshop series had noted, low-wage workers have been hit hardest by the COVID-19 pandemic, and, as Shevlin pointed out, higher-income workers have largely recovered already. The pandemic has fundamentally changed the U.S. economy, including consumer demand and the way businesses operate, with the result that there could be a permanent loss of many low-wage jobs. In fact, said Shevlin, a recent study estimated that between 32 and 42 percent of pandemic-induced layoffs will be permanent (Barrero, Bloom, and Davis, 2020). In addition, 10.1 million jobs have vanished since February 2020, so there are fewer openings for more unemployed workers to fill.

An up-side to that situation, if there is one, is that this is a good time to pursue education and training for the jobs of the future, although Shevlin pointed out that there are significant barriers to doing so, including

- Education and training programs can be expensive, and many people, especially low-wage workers, cannot afford them.
- People do not have the information and support to make informed decisions about training opportunities and career pathways.
- Other constraints, such as childcare and transportation, create additional barriers.
- Workforce and education services are not delivered equitably, exacerbating disparities along the lines of race and gender.
- Virtual and online learning may be the only feasible option for some people; however, many programs are only offered in person, and online learning presents unique challenges.

In terms of what states, regions, and local workforce boards can do without policy changes, Shevlin said they can identify effective training programs, including virtual options, and make that information widely available to address the information barrier. She pointed to the value of sending letters to unemployment insurance recipients about suitable training programs in which they could participate. During the 2008 recession, that simple step increased enrollment in postsecondary education by an astounding 40 percent, said Shevlin (Barr and Turner, 2018). She also suggested tracking program participation and setting targets for subpopulations based on the local unemployment situation.

Shevlin's organization, the Markle Foundation, along with more than 30 partners, recently launched the Rework America Alliance with the goal of delivering this type of information to organizers that serve job seekers. It has also released a set of policy recommendations aimed at giving workers the tools they need to secure a good job (Markle Foundation, 2020). The recommendations include to

- Create an Opportunity Account, a new funding system that makes effective education and training affordable for all workers to enroll in programs that get people into well-paying jobs.
- Provide matching funding to employers that will hire and train for quality jobs, creating an incentive for employers to provide input to training programs.

- Expand career coaching to provide workers with the support to navigate the job market, training options, and pathways to economic security.
- Scale and create online learning opportunities.

Shevlin noted in closing that the Rework America Alliance also aims to help millions of workers move into good jobs, regardless of their formal education, by accelerating the development of a new system of worker training aligned to quality jobs that employers need to fill.

Discussion

Rowley discussed how her institution knew almost immediately that broadband access was going to be a problem during the COVID-19 pandemic, particularly for areas such as the pueblos where there is no access at all and wireless internet hotspots will not do any good. Her team created an emergency fund that quickly raised $100,000 to support student needs for everything except tuition, including paying for internet access and purchasing laptops that students could borrow.

Jacoby noted that community colleges are a mixed bag of institutions, some of which are focused on workforce training, others on getting students into bachelor's degree–granting institutions. In general, the graduation rate is low, and the number of transferring students who go on to get their bachelor's degree is even lower. In her opinion, community colleges need to focus more on people who have lost their jobs and helping them get the retraining they need to obtain better-paying jobs. The good news here, she said, is that community colleges appear to be getting that message and are refocusing their efforts on retraining and upskilling.

When asked about the Opportunity Account that the Markle Foundation recommends, Shevlin gave an example of how displaced workers can use those funds. Apprenticeship-readiness programs, typically full-time for 12 weeks, require a trainee to forgo earning income during the day, but they still need money to support themselves, which is a major barrier to enrolling people. These Opportunity Accounts could be used to pay for gas, food, and other necessities while enrolled in such a program. "We have to find a way to provide some additional funding for people to help them live while they are in these training programs or else we are never going to see the participation that we want among the most disadvantaged, unemployed workers," said Shevlin.

Jacoby was asked to discuss a statewide initiative in Kentucky designed to train advanced-manufacturing technicians, the people who repair assembly lines in manufacturing plants, an initiative that has produced significant, positive results. She explained that this initiative, started by Toyota, now includes nearly 400 companies in 14 states operating together as the Federation for Advanced Manufacturing Education. In Kentucky, people who participated in this program are earning $45,000 a year more, five years later, than non-participating students from the same community colleges. "While not all apprenticeship programs are going to pay off in that way, this is an example of what can be done," said Jacoby. Her hope is that by studying programs such as this, researchers will identify the elements that make it a gold standard for apprenticeship programs.

To increase connectivity between employers who have jobs available and the institutions looking to train those potential employees, Shevlin suggested giving money directly to employers and having them approach the community college with funding in hand to help the community college develop the appropriate training programs. "Giving funding to the employers and letting them drive that demand is a strategy that we should at least try," she said. Her one caveat was that funding should only go to employers with high-quality jobs, and not, as is often the case with the current workforce system, to companies offering low-wage work. "Our proposal would be that we do not provide any subsidy for training for jobs that pay less than $35,000," she added.

Jacoby said that she was skeptical about that idea because she observed some employers become experts at securing this money. Shevlin countered that small and medium-sized businesses have a hard time financing training, and those are the companies whose available jobs the program seeks to support. "If you do not provide funding for training, then employers will hire people who already have the skills for the job and we are never going to address the equity gap," replied Shevlin.

One issue that Santa Fe Community College encounters is that when it initiates a workforce training program, there often is not funding available for students to enroll in that program; the institution should create that funding stream, either through grants or from an employer. "It is incumbent on us as the community college to figure out how to help a student pay for that program," said Rowley. In her mind, it is wrong to create a strong vehicle to move people out of poverty but have that vehicle be so expensive that the targeted population cannot afford to use it.

Jacoby seconded the idea of putting money in the hands of learners to obtain training, but that should be accompanied by navigational supports to help learners get into the program that is right for them and, more importantly, complete that program. She pointed out that many dislocated workers have not been in a classroom in decades, and when they were students they may not have been good ones, which may be partly why they ended up in low-wage jobs. "This is a place where the community college and workforce system should work together," said Jacoby. Shevlin agreed that this type of coaching is essential, and the Markle Foundation has a proposal to significantly expand coaching through the federal workforce system and local organizations.

Rowley noted that Santa Fe Community College is moving to incorporate coaching into its programs, and to also provide more information about the labor market and what it takes to enter a particular profession in the Santa Fe region. Such information will help both learners and coaches. Her institution is also starting to take a case management approach to helping students address academic barriers that keep them from persisting in their training.

Both Jacoby and Rowley called for making Pell grants available for short-term continuing education. Shevlin disagreed, though; since there is no accountability or transparency requirement for Pell grants, she worries that this would merely increase the amount of money going to the many short-term programs that are of poor quality. "What we need to do is separate out the low quality from the high-quality programs and then generously fund the high-quality programs so that we can make sure that we are spending public dollars well," said Shevlin.

FUTURE DESIGN OF WORKFORCE TRAINING

The final session of the workshop series examined how the COVID-19 pandemic's effect on the nature of work will drive new approaches to workforce training. The three panelists were David Langdon, senior economist and senior policy advisor in the Office of the Chief Economist at the Department of Commerce's Economics and Statistics Administration; Paul Osterman, Nanyang Technological University professor of human resources and management and professor of urban planning at the Massachusetts Institute of Technology Sloan School of Management; and Unique Morris-Hughes, the director of the Washington, DC, Department of Employment Services. Rya Conrad-Bradshaw, senior director of employer engagement at Grads of Life, moderated the session.

Langdon opened the session by discussing how the nature of unemployment is changing rapidly, shifting from large-scale and temporary job loss to more permanent job losses. Statistics from the Bureau of Labor Statistics also show that, contrary to the common perception that a great deal of employment is virtual these days, only one-fifth of employed people were teleworking because of the pandemic in October 2020. In Langdon's view, one of the main lessons to learn from those businesses and employees that are recovering is the need to be flexible and agile. In some cases, he said, businesses are finding it easier to access and hire more diverse talent in this remote-working environment. The challenge will be blending remote work and essential in-person work in a way that does not exacerbate inequities in the workforce.

Osterman shared data on how people work in terms of their relationship to their employers (Osterman, 2020). So-called gig workers represent a mere 1.0 percent of the workforce, so, said Osterman, the U.S. economy is not being "Uberized." He also noted that in-demand skills are within the reach of most people, and most community colleges can provide them. Demographics, he added, will create opportunities for middle-skill work. While the number of production worker jobs is projected to fall by nearly 430,000 positions between 2018 and 2028, there will be over 1.6 million openings during that time if current production workers retire at age 65, and 620,000 openings if all production workers wait until they are 70 to retire.

The United States, said Osterman, has a distinctive and remarkably complex job training system. There are school-based programs, firm-based programs, school-based adult lifetime learning programs, and remedial and sectoral programs. The main critique of the current system, he said, is that it is needlessly complicated, poorly articulated, and hard to navigate. The U.S. government, for example, has some 250 job training programs. On the other hand, the system is open and flexible, and it allows people to change their minds, which is hard to do in European systems, for example.

In the wake of the COVID-19 pandemic, demands on and expectations for the workforce training system are growing and will continue to grow substantially. Currently, employers are the largest source of training in the country, with more than half of employees having received employer-provided training in the previous 12 months, although employer training for contractors and first-job freelancers was less common (Osterman, 2020). Approximately 56 percent of white employees, 55.3 percent of Black employees, and 47.4 percent of Hispanic employees received on-the-job training from employers. There is a sharp drop-off associated with educational level, however, with 61.1 percent of those with a college degree and 50.1 percent of those with a high school degree receiving employer training, compared with 19.5 percent of those with less than a high school degree. These disparities, said Osterman, argue for a strong public workforce training system.

Community colleges are the main source of formal training in the public higher education system, accounting for some 12 million learners a year, about half of whom are in vocationally oriented programs. Given that the quality of the programs run at the approximately 1,200 community colleges in the United States is highly variable, determining and disseminating best practices is an essential activity going forward. Osterman cited data from a randomized controlled study of a high-quality workforce development program in San Antonio known as Project QUEST showing that earning gains for those who complete that program exceeded $5,000 at six years post-completion compared to those who went through a standard program (Elliott and Roder, 2017).

Building strong programs requires money, with most of that coming via regional compacts that include employers, workforce boards, community colleges, intermediaries such as national and regional support networks, and community actors. The hardest part, Osterman said, is involving employers and convincing governors that it is worth their political capital to establish these compacts. He noted that federal funding of job training programs has fallen since the early 2000s and, while credentialing sounds like a reasonable idea, he is not sure that employers will actually use credentials, as opposed to a formal degree, as a criterion for hiring.

The final panelist, Morris-Hughes, noted that she was speaking from the perspective of an on-the-ground practitioner whose goal is to make sure that residents of the District of Columbia have a fair shot at economic prosperity and membership in the middle class through thoughtful, strategic, and innovative programming. At the time of the workshop, her department was serving three groups of people: those who were unemployed prior to the COVID-19 pandemic, those who lost their jobs because of the pandemic, and a group that is often forgotten, the chronically underemployed or unemployed. The main challenge in reaching these three groups is the digital divide, she said, though the oversupply of individuals coming from neighboring areas of Maryland and Virginia is crowding out opportunities for District of Columbia residents. In addition, a recent survey suggests that many programs offered in the District of Columbia may not support the city's most vulnerable residents, with over 50 percent of programs reporting that they have educational requirements for entry despite targeting low-income residents.

Going forward, Morris-Hughes said that real-time jobs data over an extended time period will be crucial for ensuring that workforce training opportunities are in sync with the skills most in demand. Her department is currently building out occupational pathway maps in the context of industry pathways in order to identify programmatic training requirements and is completing a training program assessment by occupation to pinpoint any gaps and start new programs to fill those gaps. It is also conducting an assessment to identify gaps in the most critical service support areas.

Discussion

On the topic of workforce trends accelerated by the COVID-19 pandemic, Osterman shared that the need to transition people from one set of occupations to another has increased. While employment services are supposed to help with that transition, they are severely underfunded, he said, at under $100 per client, and they are poorly connected to the workforce training system. He also pointed to the importance of better understanding the type of learning opportunities that work best for a middle-aged person who may have a high school education and been out of school for decades. Morris-Hughes said that the need to address the middle-skills gap has grown substantially during the pandemic, as has the competition for those jobs in the District of Columbia. Langdon noted the need to focus on skills-based hiring and the skills-based labor market.

When asked what advice they would give to industries and regions as they think about developing their future strategies for engaging employers, Morris-Hughes said that because of COVID-19, employers and industries need a certain level of flexibility regarding the talent pipeline that exists in a region. In addition, people seeking jobs need to be responsive to local demand and keep an open mind as to the industries in which they can apply their skills. Osterman said he would like to see employers forget about national credentialing requirements, given that most jobs are filled locally and relatively few people move to a new region to find a job. In that regard, all training programs should be regional in their skills focus.

Langdon explained that the Office of the Chief Economist at the Department of Commerce's Economics and Statistics Administration does provide federal funding to support regional economic development strategies. He noted, though, that there is a niche trend toward employers hiring people from around the country to take advantage of telework. "If that were to take place on a relatively large scale, at least for certain types of occupations, would that lead to a disconnect between a local population and economic development taking place in that area?" he asked. Conrad-Bradshaw said she believes that some regions will become remote work regions, with Tulsa, Oklahoma, and Harford, Connecticut, already considering such a strategy.

In her final question, Conrad-Bradshaw asked the panelists to comment on the role of the employer going forward. Osterman said that employers are of course central given that they are the ones with jobs, and they should become more involved in partnerships with educational institutions and local workforce development boards. Langdon believes that employers are involved and are paying for most workforce development, but he acknowledged that they should be more involved as partners with other institutions in their regions. Morris-Hughes noted the importance of engaging employers in an authentic and meaningful way and recognizing them for the work they do.

To conclude the workshop, Bingaman highlighted two factors that make the subject of the six workshops particularly important at this time. One is the enormous flux occurring in the workforce today, with the large number of people who are unemployed and the number of people who are seeking a different type of employment. A second factor, he said, is the recent election that provides opportunities for people coming into office, from President Biden down to officials elected at the local level, to put job training higher on the public agenda and provide more resources to assist with job training opportunities.

COMMON THEMES DISCUSSED ACROSS THE WORKSHOP SERIES

The workshop series sought to advance conversations across different U.S. cities on meeting their STEMM workforce needs in the wake of COVID-19. While the panel discussions were focused on individual cities' needs and resources, several themes emerged across the workshop series. These themes, compiled by this summary's rapporteurs, are presented in Box 7-1.

BOX 7-1
Common Themes Across the Six Workshops

Job training offerings and curricula

- Local employment opportunities and workforce demands inform job training offerings and curricula.
- Opportunities for students and recent graduates to gain employment, either immediately after training completion or while students undergo training, can be effective at meeting local workforce needs equitably.
- Employers can facilitate career advancement.
- Addressing workforce needs requires a consortium of employers, educational institutions, and support service organizations, and can include both well-established organizations and new innovators.
- Bridging the silos of education, career services, and the labor market is key for meeting local workforce needs.
- Funding for job training programs and education is important.
- Training programs that are free to students or provide "learn and earn" opportunities can benefit diverse students and workers.
- Institutional leaders who are adaptable and flexible can encourage innovation for new ways to train people, fill high-demand jobs, and build new types of partnerships to create the workforce of the future.

Diversity and equity in workforce training

- The COVID-19 pandemic has drawn attention to as well as exacerbated unemployment disparities affecting women and people of color.
- It is time for workforce training efforts and hiring practices to move past diversity to equity and justice.

Educational emphases: Skills versus degrees

- While four-year colleges play an important role in workforce training, community colleges are particularly well suited to respond to local workforce demands because of their close ties to local industry. Community college enrollments are down, however, and many of these institutions are under-resourced.
- It is important for employers to avoid redundant and/or unnecessary educational requirements when hiring, when feasible.

This can include not requiring or expecting job candidates to have four-year degrees for jobs requiring skills that can be obtained without degrees and not requiring credit checks.

- Students benefit by having access at higher education institutions to competency-based learning and certifications.
- Competency-based education, stackable micro-credentials, and ladders of opportunity are all key to employing people quickly without imposing overly burdensome educational requirements.

The digital divide

- Many people across the nation do not have high-speed internet subscription services or access, and a large portion of the population can access the internet only through mobile devices, seriously hindering many people's ability to access virtual training opportunities or to keep or obtain employment that requires working from home.
- Many virtual training opportunities and job application platforms do not have a mobile-ready platform.
- Many places that provide internet access, such as public libraries, coffee shops, and computer labs at job training centers, are closed or have limited access as a result of the pandemic.

Social support services

- Wrap-around services and social support services, such as childcare and transportation, are crucial for meeting regional workforce needs.
- Many people require help navigating the career system and selecting training programs.
- Even when there is internet access in a student or trainee's home, obligations and distractions may make it difficult for them to study and learn online.

Data needs

- Unemployment data must be carefully examined before being used in the design of workforce training efforts. In some regions, cities, and communities, while unemployment may be low, poverty may still be high. The need may be less for more jobs per se, and more about directing people in low-wage jobs toward training opportunities for jobs that pay a living wage.
- Communities need "real-time" labor data to help them adapt to rapidly changing workforce demands.

References

Andreason, S., P. Haley, S. Miller, and M. de Zeeuw. 2020. *The digital divide and the pandemic: Working from home and broadband and internet access.* Atlanta, GA: Federal Reserve Bank of Atlanta.

Arnold, R., J. De Sa, T. Gronniger, A. Percy, and J. Somers. 2006. *A potential influenza pandemic: Possible macroeconomic effects and policy issues.* Washington, DC: Congressional Budget Office.

Barr, A., and S. Turner. 2018. A letter and encouragement: Does information increase postsecondary enrollment of UI recipients? *American Economic Journal: Economic Policy* 10(3):42-68.

Barrero, J. M., N. Bloom, and S. J. Davis. 2020. *COVID-19 is also a reallocation shock.* Washington, DC: Brookings Institution.

BLS (Bureau of Labor Statistics). 2019a. *American time use survey: 2018 results.* Washington, DC.

BLS. 2019b. *Quarterly census of employment and wages, 2019 Q1–Q3 data.* Washington, DC.

Bradley, B., C. Rudnicki, R. Neumann, D. Restuccia, J. Reddish, and S. L. Hudson. 2018. *Building (it) Together: A framework for aligning education and jobs in greater Birmingham.* Birmingham, AL: Burning Glass Technologies and Council for Adult and Experiential Learning.

Elliott, M., and A. Roder. 2017. *Escalating gains: Project QUEST's sectoral strategy pays off.* New York: Economic Mobility Corp.

Glenn, L. 2020. How are community college students different? In *Community College Daily.* Washington, DC: American Association of Community Colleges.

Markle Foundation. 2020. *Investing in workers to drive a stronger economic opportunity for all.* New York: Markle Foundation.

Osterman, P. 2020. *Skill training for adults.* Cambridge, MA: Massachusetts Institute of Technology.

Ross, M., S. Fritzberg, J. Carpenter, and S. Martinez. 2020. *Local service corps can address unemployment and community need.* Washington, DC: Brookings Institution.

WEF (World Economic Forum). 2020. *The future of jobs report 2020.* Cologny, Switzerland. https://www.weforum.org/reports/the-future-of-jobs-report-2020.

Yuen, V. 2020. *The $78 billion community college funding shortfall.* Washington, DC: Center for American Progress.

Appendix A

Statement of Task

A committee will plan and host a series of six virtual workshops to identify immediate and near-term regional science, technology, engineering, math, and medical (STEMM) workforce needs in the wake of the COVID-19 pandemic. COVID-19 has impacted workforce demand that includes, but is not limited to, data analytics for logistics and operations, health care, manufacturing, agriculture, and IT infrastructure. The workshops will highlight promising practices that 2-year and 4-year higher education institutions, as well as non-academic training and certification programs, can utilize to respond urgently and appropriately to those needs. In this context, effectiveness of the practices will be judged, in part, on the number of learners who can be reached, especially in the skilled technical workforce. A series of five regionally specific virtual sessions will address the following questions:

- How have the current and near-term regional STEMM labor market needs changed as a result of the COVID-19 pandemic, particularly the needs for the skilled technical workforce (e.g., in IT and computing fields, health tech sectors, manufacturing)?
- To what extent can a regional profile be created that links STEMM educational and training resources with COVID-19 impacted workforce needs, including retraining and certification of professionals?

- What practices and policies can local higher education institutions and workforce training organizations adopt to respond to local industry STEMM workforce needs, both individually and as part of regional coalitions? Which effective practices and policies are replicable and scalable? What barriers, if any, exist that inhibit these programs from meeting COVID-19 shifts in regional STEMM workforce needs?
- How can workforce training and education systems address the significant and urgent needs of the underrepresented minority population, which has been especially impacted by the effects of COVID-19? Can the recent events related to the death of George Floyd raise awareness of the disenfranchisement of the nation's underrepresented minority populations that result from institutionalized racism, and therefore increase focus on the particular needs of those populations?
- What actions are needed to assist local higher education institutions, training and certification programs, industry, and policy makers at the local, state, and federal levels in fostering improved linkages between education and training resources and STEMM workforce needs at the regional level as a result of the COVID-19 pandemic?

Following the five regional sessions, the committee will convene a concluding virtual workshop, and a designated rapporteur will prepare a proceedings of the workshop series.

Appendix B

Workshop Agendas

AGENDA FOR WORKSHOP #1: BIRMINGHAM, ALABAMA

Friday, October 9, 2020
12:00–3:30 PM Central Time
Webcast available at https://bit.ly/3iHFMrm

Workshop Objective

To explore the resilience, alignments, and adaptability of key workforce training institutions within Birmingham in response to the COVID-19 pandemic.

12:00–12:10 PM Welcome and workshop overview

Planning committee members welcome attendees and provides background on the workshop series.

- Dr. Josh Carpenter, Director of Economic Development for the City of Birmingham
- Maria Flynn, President and Chief Executive Officer of Jobs for the Future

12:10–12:30 PM Birmingham data

Data are provided to frame the workshop discussions, including economic, labor force, educational attainment, and demographic data for the Birmingham metro area.

- Dr. Stuart Andreason, Director of the Center for Workforce and Economic Opportunity at Federal Reserve Bank of Atlanta
- Dr. Josh Carpenter, Director of Economic Development for the City of Birmingham

12:30–1:15 PM Panel 1: Birmingham's economic outlook prior to and during COVID

Panelists discuss pre-COVID investments in workforce training programs and the impact of the pandemic on their workforce alignment and learner access.

- Mr. Joseph Parilla, Metropolitan Policy Program Fellow of the Brookings Institution
- Dr. Sanjay Singh, Advisor and Board Member of Pack Health
- Moderator: Antiqua Cleggett, Executive Director of Central Six AlabamaWorks!

1:15–1:30 PM Break

1:30–2:15 PM Panel 2: Workforce programming changes and options in Birmingham

Panelists discuss emerging workforce training programs for the Birmingham region and efforts to address labor inequities exacerbated by the pandemic.

- Suzanna Fritzberg, Executive Director of Birmingham Strong
- Rachel Harmon, Chief Executive Officer of Birmingham Promise
- Waymond Jackson, Chief Executive Officer of Ed Farm
- Emily Wykle, Acting PI and Board Member of Innovate Birmingham, Director of External Affairs, University of Alabama at Birmingham
- Moderator: LaChaun Banks, Associate Director of the Ash Center's Innovations in Government Program, Harvard Kennedy School

2:15–3:00 PM Panel 3: Higher education programming for the Birmingham workforce

Panelists discuss higher education efforts in the Birmingham region, both during the pandemic and in the future, to support local workforce development.

- Dr. Jarralynne Agee, Provost and Senior Vice President of Academic Affairs at Miles College
- Dr. Cynthia Anthony, Interim President of Lawson State Community College
- Dr. Pam Benoit, Senior Vice President of Academic Affairs and Provost at the University of Alabama at Birmingham
- Guin Robinson, Associate Dean of Economic Development at Jefferson State Community College
- Dr. Daniel Coleman, President of Birmingham Southern College
- Moderator: Cassius Johnson, Senior Director of Education Strategy Group

3:00–3:30 PM Workshop summary

Summary remarks are offered of the key themes, challenges, and policy opportunities around workforce training and program development for Birmingham.

- Dr. Josh Carpenter, Director of Economic Development for the City of Birmingham
- Sarah McMillan, Director of Workforce and Talent Development for the City of Birmingham

AGENDA FOR WORKSHOP #2: BOSTON, MASSACHUSETTS

Wednesday, October 14, 2020
12:30–4:30 PM Eastern Time
Webcast available at https://bit.ly/3lDksVu

***Workshop Objective*:**

To understand how the COVID-19 pandemic is affecting STEMM workforce challenges and opportunities in Boston, and approaches that are building on ramps with equity to address the needs of Boston workers hit hardest by this crisis.

12:30–12:40 PM Welcome and workshop overview

Planning committee members welcome attendees and provide background on the workshop series.

- Rachel Lipson, Director of the Project on Workforce, Harvard Kennedy School
- Lee Wellington, Founding Executive Director, Urban Manufacturing Alliance

12:40–1:00 PM Boston data

Data are provided to frame the workshop discussions, including economic, labor force, educational attainment, and demographic data for the Boston metro area.

- Rachel Lipson, Director of the Project on Workforce, Harvard Kennedy School

1:00–1:15 PM A recovery with equity: View from the city of Boston

This framing discussion will center around racial equity in our collective exploration of STEMM workforce training needs in the Boston region, and how those needs have shifted in the wake of COVID-19 (public official remarks).

- Trinh Nguyen, Director of Workforce Development for the City of Boston

1:15–2:00 PM Panel 1: COVID response and workforce transitions in greater Boston

This session will highlight how workforce development organizations, educational institutions, and training providers have shifted their operations and program delivery mechanisms in this COVID environment. Topics to be discussed include emerging STEMM workforce development opportunities in greater Boston, partnerships to support operations in a resource-constrained environment, and continuity in training and wrap-around services as programs shift to remote operation.

- Kaitlyn Bean, Senior Program Officer, Boston Foundation
- Lane Glenn, President of Northern Essex Community College
- Joanne Pokaski, Senior Director of Workforce Development at Beth Israel Deaconess Medical Center
- Jerry Rubin, President and Chief Executive Officer of Jewish Vocational Service Boston
- Moderator: Deeneaus "D" Polk, Bosch Fellow at the Robert Bosch Foundation

2:00–2:15 PM **Break**

2:15–3:00 PM **Panel 2: COVID and Boston's economic divide: The intersection of the pandemic, racial and socioeconomic inequality, and access to good jobs in STEMM**

This panel will focus on the ways that COVID has exacerbated existing inequities in the city, and how multi-stakeholder approaches to STEMM workforce development can directly address those inequities. The panel will explore how different institutional actors are working to embed anti-racism in their training and pipeline development activities. It will also explore how these institutions have adapted their programming in response to the death of George Floyd and the racial uprisings across our nation that followed.

- Tyra Anderson Montina, Boston Site Director of Year Up
- David Delmar Senties, Founder and Executive Director of Resilient Coders
- Valerie Roberson, President of Roxbury Community College
- Neil Sullivan, Executive Director of the Boston Private Industry Council
- Moderator: Peter Blair, Assistant Professor of Education at the Harvard Graduate School of Education

3:00–3:45 PM **Panel 3: Catalyzing STEMM entrepreneurship for economic mobility**

This panel will explore the linkages between entrepreneurship and STEMM workforce development, articulating how entrepreneurship can provide new routes to economic mobility for Bostonians coming out from this crisis. Specific topics include how entrepreneurship can create job transition pathways, provide insights into emerging STEMM careers, catalyze ownership and employment opportunities, and provide a foundation for workforce education. This discussion will also highlight how entrepreneurial support organizations—including incubators, accelerators, and makerspaces—have worked closely with community-based partners to build ecosystems with strong connections to STEMM workforce development programming, and ways this has changed as a result of the pandemic.

- Jen Faigel, Executive Director and Co-Founder of CommonWealth Kitchen
- Sonia Moin, Senior Director of Urban Business Initiatives at the Initiative for a Competitive Inner City
- Michelle Weise, Senior Advisor and Entrepreneur-in-Residence at Imaginable Futures
- Moderator: Cecilia Wessinger, Principal and Founder, Mass Collaboration

3:45–4:15PM Workshop reflections: A discussion

In this session, discussants reflect on the lessons learned from the day's discussions, with an eye to highlighting important policy implications for the labor market recovery in Boston and in the nation.

- Maria Flynn, President and Chief Executive Officer, Jobs for the Future
- Dr. Robert Schwartz, Professor Emeritus of Education at the Harvard Graduate School of Education

4:15–4:30PM Closing remarks

Closing remarks will summarize themes and takeaways from the workshop and key priorities for greater Boston's workforce ecosystem and policy, business, education, and training institutions.

- Christine Abrams, President and Chief Executive Officer, Commonwealth Corporation

AGENDA FOR WORKSHOP #3: RICHMOND, VIRGINIA

Monday, October 19, 2020
1:00–4:15PM Eastern Time
Webcast available at https://bit.ly/31bvGZD

***Workshop Objectives*:**

- *Explore the economic and STEMM-related workforce impacts from COVID-19 in Richmond, Virginia, and compare them to statewide impacts*
- *Identify promising education and workforce practices at the city and state level that are responsive to workforce disruptions brought about by COVID-19*
- *Highlight opportunities and synergies between city, regional, and state policies*

1:00–1:15 PM Welcome and workshop overview

Planning committee members welcome attendees and provide background on the workshop series.

- Dr. Stuart Andreason, Director, Center for Workforce and Economic Opportunity at the Federal Reserve Bank of Atlanta; and National Academies of Sciences, Engineering, and Medicine Task Force Committee Member
- Alastair Fitzpayne, Executive Director, Future of Work Initiative at the Aspen Institute

1:15–2:05 PM Panel 1: Understanding the economic and workforce impacts of COVID-19 on the capital and the commonwealth

This panel will frame the workshop discussions by sharing economic, labor force, education, and demographic data for the Richmond metro area, and how it compares to the state of Virginia overall. Responses will be offered by statewide and regional experts.

- Dr. Stuart Andreason, Director, Center for Workforce and Economic Opportunity at the Federal Reserve Bank of Atlanta; and National Academies of Sciences, Engineering, and Medicine Task Force Committee Member

- Leonard Sledge, Director, Department of Economic Development, City of Richmond
- Sonya Waddell, Vice President and Economist, Regional and Community Analysis, Federal Reserve Bank of Richmond
- Moderator: Mr. Alastair Fitzpayne, Executive Director, Future of Work Initiative at the Aspen Institute

2:05–3:00 PM Panel 2: How Richmond's education and training infrastructure is responding to the COVID-19 crisis

Panelists will discuss the impact of the pandemic on workforce education and training programs for the Richmond region.

- Dr. Paula Pando, President, Reynolds Community College
- Dr. Hakim Lucas, President, Virginia Union University
- Dr. Peter Buckley, Dean, Virginia Commonwealth University School of Medicine
- Brian Davis, Executive Director, Capital Region Workforce Development Board
- Mark Barth, President and Chief Executive Officer, Goodwill Central and Coastal Virginia
- Moderator: Jane Oates, President, Working Nation; Former Assistant Secretary for the Employment and Training Administration, Department of Labor

3:00–3:10 PM Break

3:10–4:00 PM Panel 3: Innovative workforce solutions: How state and local policy and programs are meeting the COVID-19 challenge

Panelists will discuss post-COVID policy responses and investments in workforce training programs, including the impact of the pandemic on workforce alignment and learner access.

- Dr. Megan Healy, Chief Workforce Advisor to Virginia Governor Ralph Northam
- Valaryee Mitchell, Director, Office of Community Wealth Building, City of Richmond

- Caren Merrick, Chief Executive Officer, Virginia Ready Initiative
- Elizabeth Creamer, Vice President of Workforce Development and Credential Attainment, Community College Workforce Alliance
- Nat Marshall, Chair, Virginia Board of Workforce Development
- Moderator: Anna Payne Fife, Associate Director for Outreach and Engagement, Aspen Institute Future of Work Initiative

4:00–4:15 PM Conclusion and workshop summary

Closing remarks will summarize the key themes, challenges, and policy opportunities related to workforce development for Richmond and the state of Virginia.

- Dr. Josh Carpenter, Director of Economic Development for the City of Birmingham, National Academies Science, Engineering, and Medicine Task Force Member

AGENDA FOR WORKSHOP #4: RIVERSIDE, CALIFORNIA

Thursday, October 22, 2020
10:00 AM–1:15 PM Pacific Time
Webcast available at https://bit.ly/2FALOwm

Workshop Objectives

- *This workshop will focus on creating equity within the workforce pipeline and connecting it to the green economy and the high tech, high growth jobs of tomorrow. Moreover, we will highlight the importance of the region's logistics industry and its valuable role in addressing the problems presented to the state and local workforce in the post-COVID labor market. In addition, we will identify training opportunities to address challenges and build a more resilient workforce; identify challenges presented to the workforce on a regional, state, and local level post-COVID; and identify upskilling and reskilling opportunities at local universities and community colleges to connect the transitioning workforce into the green economy and high tech, high-demand careers.*

10:00–10:10 AM Welcome and workshop overview

Planning committee members welcome attendees and provide background on the workshop series.

- Van Freeman, Director for Future Workforce Development and Strategy, Aerospace Industries Association
- Chancellor Eloy Ortiz Oakley, Chancellor, California Community Colleges

10:10–10:30 AM Riverside data

Data are provided to frame the workshop discussions, including economic, labor force, educational attainment, and demographic data for the Riverside region.

- Dr. Stuart Andreason, Director of the Center for Workforce and Economic Opportunity at Federal Reserve Bank of Atlanta

10:30–11:15 AM Panel 1: Greater Riverside economic outlook prior to and during COVID

Panelists will discuss pre-COVID investments in workforce training programs and the impact of the pandemic on their workforce alignment and learner access.

- Dr. Kim Wilcox, Chancellor, University of California, Riverside
- Toni Symonds, Chief Consultant to the Committee on Jobs, Economic Development, and the Economy, California State Assembly
- Dr. Wolde-Ab Isaac, Chancellor, Riverside Community College District
- Moderator: Chancellor Eloy Ortiz Oakley, Chancellor, California Community Colleges

11:15–11:30 AM Break

11:30 AM–12:15 PM Panel 2: Regional ecosystems and opportunities

This panel will focus on emerging workforce training programs for the Riverside region, including apprenticeships and credentialing programs.

- Dr. Noel Hacegaba, Deputy Executive Director, Port of Long Beach
- Leslie Trainor, Deputy Director of Workforce Development, County of Riverside
- Chris Cagle, Regional Affairs Manager, South Bay Workforce Investment Board
- Professor Rex Beck, Professor of Business Administration, Norco College
- Moderator: Sid Voorakkara, Senior Vice President, Strategies 360

12:15–1:00 PM Panel 3: Higher education's role in economic mobility

This panel will explore current and future efforts by higher education institutions to support local workforce development, with particular attention to increasing the diversity of the STEMM workforce pipeline and tech transfer and entrepreneurship growth and expansion.

- Dr. Jeannie Kim, Vice Chancellor of Educational Services, Riverside Community College District
- Dr. Rodolfo Torres, Vice Chancellor for Research and Economic Development, University of California, Riverside
- Dr. Byron Ford, Associate Dean, Pre-Clerkship Medical Education, University of California-Riverside School of Medicine
- Moderator: Van Freeman, Director for Future Workforce Development and Strategy, Aerospace Industries Association

1:00–1:15PM Workshop summary: Policy implications

In this closing session, discussants will summarize the lessons learned from the day's discussions, with an eye to highlighting key policy implications for the workforce recovery in Riverside and in the nation.

- Dr. Wolde-Ab Isaac, Chancellor, Riverside Community College District

AGENDA FOR WORKSHOP #5: WICHITA, KANSAS

Friday, October 30, 2020
12:00–3:15 PM Central Time
Webcast available at https://bit.ly/33TrCim

Workshop Objective

To understand how the COVID-19 pandemic is affecting STEMM workforce challenges and opportunities in Wichita, and explore approaches that are building on ramps with equity to address the needs of Wichita workers hit hardest by this crisis.

12:00–12:10 PM Welcome and workshop overview

Planning committee members welcome attendees and provide background on the workshop series.

- Karen Elzey, Associate Executive Director of Workcred
- Dr. Donna Ginther, Roy A. Roberts Distinguished Professor of Economics and Director of the Institute for Policy and Social Research at the University of Kansas

12:10–12:30 PM Understanding the economic and workforce impacts of COVID-19 on Wichita

Data are provided to frame the workshop discussions, including economic, labor force, educational attainment, and demographic data for the Wichita metro area.

- Dr. Donna Ginther, Roy A. Roberts Distinguished Professor of Economics and Director of the Institute for Policy and Social Research at the University of Kansas

12:30–1:15 PM Panel 1: Understanding state and local policy responses

This panel will discuss how COVID-19 has shifted Wichita's labor and workforce efforts and strategies.

- Scot Rigby, Assistant City Manager and Director of Development Services, City of Wichita
- Amanda Duncan, Vice President and Chief Business Development Officer, Workforce Alliance of South Central Kansas
- Trent Armbrust, Director of Strategy for Bioscience and Technology, Kansas Department of Commerce
- Moderator: Deb Miller, Director of the Public Management Center, University of Kansas

1:15–1:30 PM Break

1:30–2:15 PM Panel 2: How Wichita's higher education workforce training efforts are responding to the COVID-19 crisis

This session will highlight changes in higher education workforce development programs as a result of COVID for the Wichita region.

- Dr. Susan Norton, Director of Adult Learning, Wichita State University
- Kimberly Moore, Director for Workforce, Professional, and Community Education, Wichita State University
- Christopher Stanyer, Vice President of Career Services, Goodwill Industries of Kansas
- Deltha Colvin, Associate Vice President for Special Programs in the Office of Regional Engagement and Economic Development, Wichita State University
- Moderator: Karen Elzey, Associate Executive Director of Workcred

2:15–3:00 PM Panel 3: How Wichita's workforce training infrastructure is responding to the COVID-19 crisis

Panelists will discuss post-COVID policy responses and investments in workforce training programs, including the impact of the pandemic on workforce alignment and learner access.

- Michele Gifford, Human Resources Director for Operations and Workforce Programs, Textron Aviation
- Amy Williams, Senior Manager for Global Community Relations and Corporate Administration, Spirit AeroSystems

- Cheryl Childers, Human Resource Manager, Cox Machine
- Moderator: Dr. Lida Beninson, Senior Program Officer, National Academies of Sciences, Engineering, and Medicine

3:00–3:15 PM Closing remarks

Closing remarks will summarize the themes and takeaways from the workshop and key priorities for Wichita's workforce ecosystem and policy, business, education, and training institutions.

- Dr. Lida Beninson, Senior Program Officer, National Academies of Sciences, Engineering, and Medicine

AGENDA FOR WORKSHOP #6:

Regional Observances and a National Perspective

Tuesday, November 10, 2020
1:00–3:30PM Eastern Time
Webcast available at https://bit.ly/364gtuY

Workshop Objective

To summarize the adaptability, challenges, and needs of STEMM workforce training efforts across the nation in the wake of the COVID-19 pandemic.

1:00–1:10 PM Welcome and workshop overview

The committee chair welcomes attendees and provides background on the workshop series.

- Chair Jeff Bingaman, Former U.S. Senator for New Mexico

1:10–1:45 PM Regional STEMM workforce training responses in the wake of COVID-19

In this session, committee members will discuss the common themes and key differences across the five cities highlighted in the workshop: Birmingham, Alabama; Boston, Massachusetts; Richmond, Virginia; Riverside, California; and Wichita, Kansas.

- Maria Flynn, President and Chief Executive Officer of Jobs for the Future
- Dr. Donna Ginther, Roy A. Roberts Distinguished Professor of Economics and Director of the Institute for Policy and Social Research at the University of Kansas
- Rachel Lipson, Director of the Project on Workforce, Harvard Kennedy School
- Moderator: Karen Elzey, Associate Executive Director of Workcred

1:45–2:00 PM National labor, education, demographic, and broadband access data

Data are presented to frame the workshop discussions on national efforts to address training needs for STEMM jobs.

2:00–2:45 PM Panel 1: Digital infrastructure to equitably support workforce training

Panelists discuss efforts to improve nationwide broadband access so displaced workers can participate in virtual workforce training programs.

- Matt Dunne, Founder and Executive Director of the Center on Rural Innovation
- Laura Spining, Director for Broadband Infrastructure at the Department of Commerce's National Telecommunications and Information Administration
- Nicol Turner Lee, Senior Fellow in Governance Studies and Director of the Center for Technology Innovation at the Brookings Institution
- Moderator: Rick Seltzer, Projects Editor at Inside Higher Ed

2:45–3:30 PM Panel 2: Key services that support equitable access to workforce training

Panelists discuss higher education and social service strategies to connect displaced workers with training opportunities to meet local STEMM employment needs.

- Tamar Jacoby, President and Chief Executive Officer of Opportunity America
- Rebecca Rowley, President of Santa Fe Community College

- Paige Shevlin, Director of Policy and National Initiatives at the Markle Foundation.
- Moderator: Scott Gullick, Senior Director of Nest, YearUp's Research and Design Laboratory

3:30–4:15 PM Panel 3: Future design of workforce training

Panelists discuss how the pandemic's impact on the nature of work will drive new approaches to workforce training.

- David Langdon, Senior Economist and Senior Policy Advisor in the Office of the Chief Economist at the Department of Commerce's Economics and Statistics Administration
- Paul Osterman, Nanyang Technological University Professor of Human Resources and Management and Professor of Urban Planning at the Massachusetts Institute of Technology Sloan School of Management
- Unique Morris-Hughes, Director of the Washington, DC–based Department of Employment Services
- Moderator: Rya Conrad-Bradshaw, Senior Director of Employer Engagement at Grads of Life

4:15–4:30 PM Workshop summary

Closing remarks will summarize the key takeaways from the workshop series.

Appendix C

Planning Committee Biographies

Jeff Bingaman is a former United States Senator from New Mexico, serving from 1983 to 2013. He served as Chairman of the Committee on Outreach for the Senate Democratic Caucus. Senator Bingaman held several committee assignments during his tenure in the U.S. Senate, including the Committee on Energy and Natural Resources, Committee on Finance, Joint Economic Committee, Committee on Armed Services, and the Committee on Health, Education, Labor, and Pensions. On the Senate Energy Committee, he contributed to every major piece of energy policy legislation over two decades. Prior to serving as a U.S. Senator, he worked as a private practice attorney, served as counsel to the New Mexico Constitutional Convention of 1969, and was Attorney General of New Mexico from 1979 to 1983.

Stuart Andreason is the director of the Center for Workforce and Economic Opportunity at the Federal Reserve Bank of Atlanta. He conducts research and works across the country to support Federal Reserve and partner organization efforts in workforce development, the labor market, and economic opportunities for low- and moderate-income workers. Andreason has been at the Federal Reserve since 2014 and previously served as a senior adviser on human capital and workforce development. In that role he has published articles on workforce development practice and policy and labor market trends, including deep analysis of opportunity occupations, or middle-skill jobs that pay high wages. He is the editor of *Developing Career-*

Based Training and Models for Labor Market Intermediaries. Prior to joining the Atlanta Fed, Andreason was a research associate at the Penn Institute for Urban Research at the University of Pennsylvania. There, he helped develop a set of indicators of livable and sustainable communities for the U.S. Department of Housing and Urban Development funded by the Ford Foundation. He was a fellow of the Department of Education's Institute of Education Sciences at Penn and a Lincoln Institute of Land Policy C. Lowell Harriss fellow. Previously, he led two nonprofit organizations focused on economic revitalization in central Virginia and worked for the Pew Partnership for Civic Change. He has bachelor's and master's degrees in urban and environmental planning from the University of Virginia and a Ph.D. in city and regional planning from the University of Pennsylvania.

Josh Carpenter served as the Director of Economic Development for the City of Birmingham. He leads a team focused on executing a vision of making Birmingham a hub of qualified and diverse talent and a premier destination for small businesses, startups, and businesses looking to expand, propelling shared prosperity through innovation and inclusive growth. Previously, Josh served as the Director of External Affairs in the Office of the President at the University of Alabama at Birmingham where he worked on strategic projects in a portfolio of issues in community engagement and economic development. In 2016, Josh spearheaded a partnership between 15 partners and 30 employer sponsors to secure a $6 million grant from the U.S. Department of Labor, and he led a team to train 925 disconnected youth for high-demand, high-wage IT jobs in Birmingham. A Rhodes Scholar, Josh earned his M.Sc. in comparative social policy and a D.Phil. in politics at the University of Oxford where his dissertation explored the intersections of the political participation and healthcare access. Josh was named a young Global Shaper by the World Economic Forum, and he is currently a member of the Aspen Institute's Inclusive Innovation in American Cities Collaborative. Josh teaches in the UAB Collat School of Business and researches how job attainment alters notions of efficacy.

Karen Elzey is associate executive director at Workcred, an affiliate of the American National Standards Institute (ANSI), where she advances Workcred's mission to strengthen workforce quality by improving the credentialing system, ensuring its ongoing relevance, and preparing employers, workers, educators, and governments to use it effectively. Ms. Elzey has over 20 years of experience in areas of workforce development. Most recently, she was the vice president of the Business-Higher Education Forum (BHEF), where she was involved with incorporating 21st-century competencies—critical thinking, problem solving, analytical reasoning, communication, and working in multi-cultural teams—into the undergraduate experience to prepare students with the knowledge and experience to succeed in high-performing workplaces. She also led several of BHEF's projects, including the integration of data analytics into specific academic disciplines, the development of cybersecurity undergraduate programs, and the creation of new media engineering programs. Ms. Elzey previously served as vice president of the U.S. Chamber of Commerce's Institute for a Competitive Workforce, where she worked on K-12 education reform and job training policies. She was the founding director of Skills for America's Future, an employer-led policy initiative that was initially part of the Economic Opportunities Program at the Aspen Institute. Her previous experience also includes coordinating public-private partnerships between K-12 school districts and employers for a local economic development agency in Indiana, and teaching English as a second language in Poland. Ms. Elzey earned her bachelor's and master's degrees from Miami University in Oxford, Ohio.

Alastair Fitzpayne serves as Executive Director of the Future of Work Initiative at the Aspen Institute. Fitzpayne leads a staff focused on policy ideas at the federal, state, and local levels to improve economic security for traditional and independent workers and expand investment in and access to education and training programs. Prior to his work at the Aspen Institute, he served as Chief of Staff at the Department of Health and Human Services (HHS) under Secretary Sylvia Burwell. In addition to serving as Chief of Staff at HHS, Fitzpayne also held a number of senior roles at the Department of the Treasury during the Obama Administration, including Deputy Chief of Staff and Assistant Secretary for Legislative Affairs. He has also served on Capitol Hill, in both the Senate and the House, as an economic advisor to Sen. Evan Bayh and Rep. Rahm Emanuel, respectively. He has also worked at the White House Office of Management and Budget and Abt Associates, a public policy consulting firm. Fitzpayne holds a B.A. from Vassar College and an M.P.P. from the University of California-Berkeley.

Maria Flynn is president and CEO of Jobs for the Future (JFF), a national nonprofit that drives transformation in the American workforce and education systems. Maria is a national authority on the future of work, the role of technology in the labor market, career pathways for underserved individuals, and employer engagement. In 2018, Maria launched JFFLabs within JFF to bridge the traditional education and workforce systems with innovative approaches and technology-enabled solutions. Before becoming CEO in 2016, Maria was JFF's senior vice president and led the Building Economic Opportunity Group, helping entry-level workers advance to family-supporting careers while enabling employers to build and sustain a productive workforce. She also led JFF's federal policy and advocacy strategies, which focus on advancing the educational needs of underserved Americans and developing a skilled workforce. Before joining JFF in 2007, Maria was a member of the federal government's Senior Executive Service in the U.S. Department of Labor (DOL), where she held several high-level positions involving employment, training, and research. At the DOL's Employment and Training Administration, she oversaw the development of policies for training programs serving both young people and adults, supervised the agency's research and evaluation strategy, and managed its $12 billion annual budget. She was a key driver of the Workforce Investment Act of 1998, the National School-to-Work initiative, and the Secretary's Commission on Achieving Necessary Skills initiative. Maria received her B.A. from Saint Joseph's University and her master's of government administration from the University of Pennsylvania.

Van Freeman served as Director for Future Workforce Development and Strategy for the Aerospace Industries Association, where he is focused on developing a strong workforce pipeline for members through programs with government agencies, national STEM skills initiatives, and coherent workforce policy principles. Van previously served as Deputy Chief of Staff at the DC Department of Employment Services, directing the District of Columbia's state labor agency. He oversaw all employment, training, business development, labor law enforcement, worker safety, universal paid family leave, and unemployment insurance programs funded at more than $150 million in local, state, and federal funds. He also managed an additional $80 million in active capital projects, including the DC Infrastructure Academy and the Universal Paid Family Leave program. Freeman managed the strategic messaging, communications, legislative portfolio, stakeholder engagement, and regulatory rule-making process for the Department, as well as human resource and talent acquisition for over 700 employees and the unemployment compensation unit that provides benefits to all district residence. Van also served as the Senior Advisor to the National

Director at the Minority Business Development Agency (MBDA), where he advised the National Director and senior leadership team on strategic decisions, national programing, high level fiscal approaches, and congressional priorities. Before Freeman transitioned to his role at MBDA, President Obama appointed him to serve as a Deputy Director in the Department of Commerce's Office of the Secretary in 2013. Van began his career at General Motors (GM) as a manufacturing engineer, eventually moving up to production supervisor at GM's parts distribution center in Pontiac, MI. He then held a position with UBS Investment Bank in London, UK, as a business analyst. He then joined President Obama's campaign staff in Cleveland, OH, for the 2008 and 2012 election cycles. Van holds a bachelor of science in mathematics from Morehouse College and juris doctorate from Southern University Law Center.

Donna Ginther is the Roy A. Roberts Distinguished Professor of Economics and Director of the Institute for Policy & Social Research at the University of Kansas and a research associate at the National Bureau of Economic Research. Prior to joining the University of Kansas faculty, she was a research economist and associate policy adviser in the regional group of the Research Department of the Federal Reserve Bank of Atlanta from 2000 to 2002, and taught at Washington University from 1997 to 2000 and Southern Methodist University from 1995 to 1997. Her major fields of study are scientific labor markets, gender differences in employment outcomes, wage inequality, scientific entrepreneurship, children's educational attainments, and child abuse and neglect. Dr. Ginther has advised the National Academy of Sciences, the National Institutes of Health, and the Sloan Foundation on the diversity and future of the scientific workforce. She was a member of the Board of Trustees of the Southern Economic Association and of the Committee on the Status of Women in the Economics Profession of the American Economic Association. Dr. Ginther received her doctorate in economics in 1995 from the University of Wisconsin-Madison.

Rachel Lipson is the inaugural Project Director of the Project on Workforce at Harvard University's Malcolm Wiener Center for Social Policy. Launched in 2019, the Project on Workforce is a cross-school, interdisciplinary Harvard initiative focused on policy and research at the intersection of education and labor markets, and is a collaboration with the Harvard Business School Managing the Future of Work Project and the Harvard Graduate

School of Education. In this role, Rachel leads the Cross-Harvard Study Group on Future of Work and spearheads practitioner-focused research and convening initiatives that bring together perspectives from policy, business, and education. Prior to joining the Wiener Center, Rachel was Director of Special Projects at Year Up, a social enterprise and workforce training provider that connects young adults without four-year degrees to meaningful careers in some of America's most successful companies. She has previously worked on strategy for JPMorgan Chase's philanthropic investments in workforce training, on small business growth and effective governance at the World Bank, on community college research in California, and in training and organizing on President Obama's re-election campaign. Rachel is an alum of Harvard College and the joint MBA/MPP program with Harvard Business School and Kennedy School of Government.

Eloy Ortiz Oakley was appointed the chancellor of the California Community Colleges in December 2016. Prior to that, he served as the Superintendent and President of Long Beach City College, a position he held from 2007 to 2016. He joined Long Beach City College in 2002, first as the vice president of administrative services then as the College's executive vice president of administrative services from 2004 to 2006. Chancellor Ortiz Oakley was vice president of college services at Oxnard College from 2001 to 2002 and an assistant vice president at Keenan and Associates from 1999 to 2001. He is best known throughout California and the nation for implementing innovative programs and policies that help students succeed in college. He strongly believes that California's emerging economies demand a workforce with quality credentials and that the state's 113 community colleges play a pivotal role in moving California forward. Under his leadership, the Long Beach Community College District has received numerous awards and recognitions for its efforts to improve student completion rates and for directly supporting a strong small business and entrepreneurship ecosystem throughout the greater Southern California region. Chancellor Ortiz Oakley earned a bachelor of arts in environmental analysis and design and a master of business administration from the University of California, Irvine. Chancellor Ortiz Oakley was appointed as a Regent in 2014 by Governor Brown to a term ending in 2024.

Lee Wellington is the Founding Executive Director of the Urban Manufacturing Alliance (UMA), a national nonprofit that builds robust, inclusive manufacturing sectors in more than 250 cities across the United States.

Through a mix of collaborative research and ecosystem building and under Lee's leadership, UMA has become nationally recognized for knowledge transfer across public agencies and community-based organizations. She has led the UMA team in building national learning communities on a range of issues including access to capital, community-embedded workforce programming, and mission driven industrial real estate development. In 2018, Lee guided UMA's flagship research project, the State of Urban Manufacturing, a six-city study on small-scale manufacturing involving multiple Federal Reserve Banks, research universities, and hundreds of local manufacturing practitioners. Lee is a frequent presenter on urban manufacturing at conferences nationally and internationally, including Berlin's Urban Tech Summit, the White House's National Week of Making, the International Business Innovation Association's Conference on Business Incubation, and the National League of Cities' City Summit. Prior to UMA, Lee worked in the public sector at the NYC Mayor's Office of Industrial and Manufacturing Businesses and the New York City Council, at nonprofit planning organizations including the Pratt Center for Community Development and the Waterfront Alliance, and at legal services organizations including the Urban Justice Center's Street Vendor Project and South Brooklyn Legal Services. Lee holds a J.D. from Brooklyn Law School, an M.S. in city and regional planning from Pratt Institute, and a B.S. in economics from the Stern School of Business at New York University.